G. B. SELDEN.
ROAD ENGINE.

No. 549,160.   Patented Nov. 5, 1895.

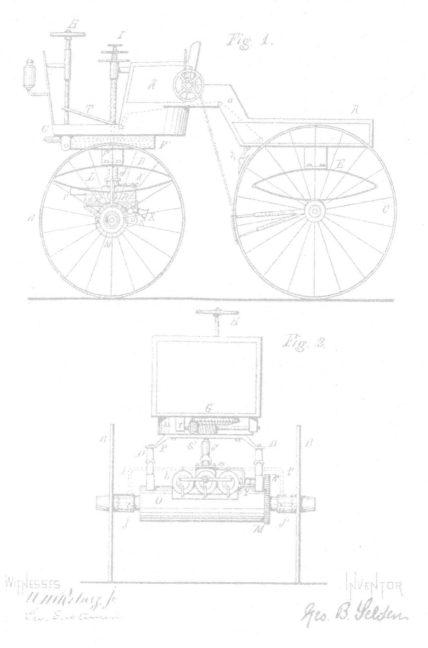

Fig. 1.

Fig. 2.

WITNESSES
H. M. Whiting, Jr.
Geo. Eastman

INVENTOR
Geo. B. Selden

W. G. MACOMBER.
ROTARY ENGINE.
APPLICATION FILED APR. 5, 1911.

1,042,018.

Patented Oct. 22, 1912.

3 SHEETS—SHEET 1.

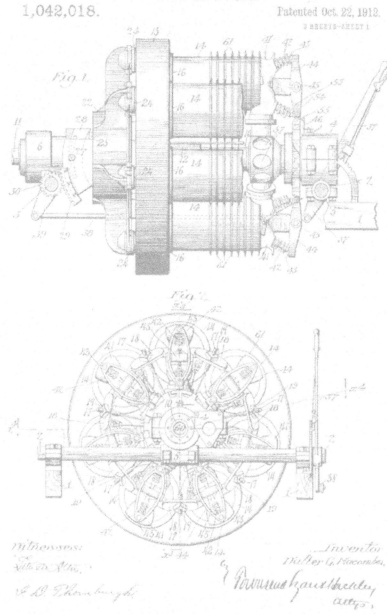

Fig.1.

Fig.2.

Inventor
Walter G. Macomber
Attys.

HOW AUTONOMOUS VEHICLES WILL CHANGE THE WORLD

ANTHONY RAYMOND

Happy the man who sees
from either aspect the glory
of these outspread wings.
The roads of his soul lie
clear, and he and his friends
shall find easy-going.

- E.M. Forster

Figure 1 - Daedalus ponders his newly constructed wings. In the background, a winding road leads up to the Labyrinth of Crete. (Modified image based on an 1895 etching of unknown origin, British Museum of London.)

This book is dedicated to the two Gils.

CONTENTS

Introduction

In a 1977 interview with the New York Times, the director and comedian Woody Allen famously quipped:

"Showing up is 80 percent of life."

This short sentence goes a long way in describing the burden that man must bear for the facility of locomotion. Even in our newly connected world—where every cellphone contains a GPS receiver and every shipping box includes a tracking code—the conveyance of people and parcels remains a logistically tedious endeavor.

Take a moment to contemplate the number of industries that directly or indirectly contribute to the transportation enterprise. Consider the amount of capital that is devoted to the chore of mere movement. Transferring your body from Point A to Point B (just "showing up") is expensive—in terms of both time and fare. So much so that the people in our lives who manage to accomplish this feat, tend to also be the ones that we buy from, or converse with, or hire, or date, or marry. Despite centuries of progress, the burden of *people-moving* is only achieved following a complex, inefficient, and error-prone choreography. However, given the fusion of several recent technological innovations, we may have reason to be optimistic about our prospects in this dance. Mankind may be about to embark on an *automotive renaissance*—an enlightened age in which he finally attains the instrument that will enable him to solve the *riddle of logistics*.

When the cost of moving a box of cheesecake across town is free, when the distance that lies between two people can be traversed without effort, and when the act of mere movement is no longer such a limiting burden, then the lives of every person on this planet should substantially improve.

Future historians may come to refer to the current segment of our journey as the "dark ages of transportation." But, thanks to the blind ambition of generations of scientists, there are finally glimmers of light on the horizon of this long road. A new type of transport will soon emerge that will redefine the way we think about mobility, distance, labor, and time.

Flying Cars?

Men have always longed for the power of omnipresence—the ability to be in all places at once. Or, in the least, the ability to traverse great distances very quickly. Historically, it has been postulated that this feat could be best accomplished via sorcery or flight.

- The 14th-century Kebra Nagast proclaimed that King Solomon owned a flying carpet—capable of carrying him to Syria for breakfast and Iran for supper.
- In the Islamic Hadith, a winged beast "the Buraq" carried the prophet Muhammad from Mecca to Jerusalem.
- In Greek mythology, the master craftsman Daedalus (whose image appears on the first page of this book) built wings so that he and his son Icarus could escape from Crete.

Contemporary narratives (especially those favored by Hollywood directors and Silicon Valley startups) have enticed us with tales of "flying cars." One of the first science fiction films ever produced (Fritz Lang's 1927 *Metropolis*), predicted that both airplanes and automobiles would casually mingle between the skyscrapers of future cities.

Figure 2 - The 1927 silent film "Metropolis" depicted a future in which both cars and airplanes weaved between skyscrapers with ease.

Improvements in cinematic special effects have allowed for more impressive depictions of speculative technology. Many sci-fi movies feature sexy metallic vehicles that swoop across the brightly lit neon skylines of the future. Some notable examples include: Blade Runner, The Fifth Element, Harry Potter, The Jetsons, and Back to the Future.

Science fiction fans and engineers alike have long scanned the horizon in anticipation of the arrival of flying cars. Yet they remain conspicuously absent from our airways.

- Back to the Future II was released in 1989. The story was set in the fictional town of Hill Valley, California in the year 2015.
- Blade Runner was released in 1982 and set in Los Angeles, California in the year 2019.

Observant readers will note that these dates have come and gone; and we never got our flying cars.

Figure 3 - Ridley Scott's 1982 film "Blade Runner" was set in a future Los Angeles of the year 2019.

This may come as a shock to some, but you will most likely never own a flying car. And nor would you really want to. Executing a vertical take-off in such vehicles would be noisy, time-consuming, and privy to the slightest variation in wind conditions. But, more importantly, for the vast majority of urban trips, achieving flight would be of little value. US motorists drive about 29.2 miles per day. Much of this driving takes place in densely populated residential areas—where flying cars would never be allowed to enter anyway. VTOL aircraft (helicopters and passenger drones) can be useful for specialized purposes—like direct-to-airport drop-offs or transportation over lakes and mountainous terrain. But independently pilotable "flying cars" (as depicted in Hollywood films) will most likely *not* be a part of your future.

Thankfully, the world of tomorrow will bring us something better (or at least more efficient) than flying cars. Unlike the CGI-generated backdrops of Hollywood, the landscape of *your* future will be a bit more down-to-earth. It will feature swarms of multi-shaped *autonomous vehicles*—darting across the streets of our highways and neighborhoods with the same agile competence that bees exhibit when they buzz across our flower gardens.

Follow the Money

Almost every daily newsfeed features at least one story about the coming age of self-driving cars and the vast amount of capital that is currently being invested into making them a reality. Aside from experimental projects by well-known self-driving car aficionados Google and Tesla, many other big names have entered the space.

- The biggest auto companies (like Ford, GM, Toyota, Volvo, and Hyundai) all threw their hats in years ago.
- So have many tech companies—like Apple, Intel, and Microsoft.
- As for ride-hailing companies (like Uber, Lyft, Via, and Juno) they indeed know that the future of their business model is entirely dependent upon their ability to seize the reigns of self-driving A.I. In April of 2019, Uber's self-driving unit (the Advanced Technologies Group) secured an impressive one billion dollar investment from the Japanese multinational conglomerate SoftBank Corp.
- Academia wants their share of the pie as well. All across the globe, thousands of university research projects have devoted millions of dollars (and classrooms full of brainiacs) to solving the self-driving riddle.

Because investment in this domain is so widespread and ongoing, it's difficult to put a price tag on it. But the current level of investment is somewhere in the 100s of billions.

Audacious timelines about the eventual deployment of self-driving technology are easy to find.

- Toyota and Honda both want their driverless cars to be utilized in the 2021 Olympics in Tokyo.

- Ford, Volvo, and Chrysler have declared that they will achieve autonomy by 2021 as well.
- The Renault-Nissan-Mitsubishi Alliance wants their vehicles to have city autonomy by 2025.

The Tesla Model 3 is purportedly already outfitted with all of the hardware necessary for autonomous driving. Given that the Model 3 was America's best-selling small and midsize luxury sedan in 2019, it appears that there may be many more *self-driving-capable* cars on the road than most people would assume.

2019 US Midsize Luxury Car Sales (Units Sold)

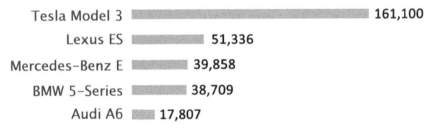

Figure 4 - Top five midsize luxury car sales of 2019. The Tesla Model 3 already includes all of the hardware (but not the software) needed for fully autonomous driving. (Source: 2019 U.S Midsize Luxury Car Sales Analysis compiled by Goodcarbadcar.net Auto Sales Data & Statistics.)

Turning your Tesla into a "fully autonomous vehicle" only requires two things to happen:

1. The vehicle's onboard computer will need an over-the-air software update.
2. The legislators in your municipality will need to take a risk— permitting these headless horsemen to share the streets with their voters.

When these steps are taken—when cities allow cars to engage their self-driving functions on public roads—then we'll know that the autonomous revolution is well underway.

Most books on futurology are forced to guess about future technological achievements—describing inventions that *might* come along someday. But, this book doesn't contain much speculative technology. The devices that we'll be discussing already exist.

- This is not a work of science fiction.
- This is not a book about "cool cars."
- This is not a hypothesis about possible future events.

Instead, this is a book about *your* future. Autonomous vehicles are coming. Indeed, if you currently own a Tesla Model 3, then you're already sitting on one.

The Four Features of Autonomy

Why are so many car companies racing to add self-driving features to their vehicles? Why are so many billions of investment dollars being poured into autonomous technology? Sure, it would be nice if your car had a button that engaged a "robot chauffeur"—a bot that could take the wheel during your morning commute. But will this technology really "change the world?"

What's so great about self-driving cars?

When futurists talk about autonomous vehicles, they're not just talking about passenger cars with an optional self-driving feature. Instead, the vehicles of the future will have four distinct attributes—all wrapped up in a single platform.

They will be:

1. Autonomous
2. Electric
3. Connected
4. Utilized as a Service

We'll briefly describe each of these features now.

Feature 1: Autonomous

Truly autonomous vehicles are more than just cars that are capable of spinning their own steering wheels. Instead, the cars of the future won't be built with steering wheels at all. Nor will they necessarily feature instrumentation or even a dashboard. From the rider's perspective, the car will only contain a seat and an interface—through which passengers can request transit to a destination address. Once this data is entered, then additional navigational input is not required of the rider. He can sit back and take a nap for the journey.

Feature 2: Electric

Given recent advances in power generation, battery manufacturing, and battery storage capacity, it is possible that every single vehicle on the road (and perhaps every motor on the planet) will be electric someday. The difficulties involved in the manufacturing of lithium-ion batteries (the "gasoline" that fuels your laptop and your Tesla) have long been the bottleneck for electric vehicle engineers. But with over a hundred battery mega-factories in the pipeline for construction, production capacity is expected to quadruple over the next decade. Additionally, electric power generation may be about to get a boost from solar. In 1970, solar power was costing us about $100 per watt. But the cost of solar has dropped by 11% each year. In 2009, the Arizona-based solar panel manufacturer *First Solar Incorporated* became the first company to tout prices of only $1 per watt (or six cents per kilowatt-hour). In September of 2017, the US Department of Energy (DOE) announced that its SunShot Initiative had achieved its goal of moving the average price of utility-scale solar to six cents per kilowatt-hour as well. The DOE hopes to cut this cost in half by 2030.

Feature 3: Connected

In the future, pretty much everything will be connected to the internet—including all wheeled vehicles. Self-driving cars will have access to real-time data (like traffic and weather reports) and they will use this information to dynamically calculate an ideal route to each passenger's destination. Along with connectivity to the city grid, self-driving cars will also be connected to each other. Currently, human drivers advertise their intentions on the road using turn signals, headlights, and angry glances. But future cars might be constantly broadcasting their position, approximate route, and projected future speed. With this information, each vehicle in the swarm of traffic will

To better describe the technological wonders that will be available to us in the world of tomorrow, let's try a narrative.

Storytime

Imagine some bright Monday morning in the year 2030 (or perhaps 2040, or 2050, or some other date that you can't hold me accountable to). You are the head of a middle-class household—tasked with getting the kids to school and getting yourself to work. As you sit at the kitchen table, typing an email on your notebook, you notice that one kid is just about ready to go while the other one is running late (as usual). A chime rings on the phone of Ethan— the more punctual child—indicating that his school bus will be stopping nearby in a couple minutes.

"Are you going to make it to the pick-up spot on time?" you ask.

"Yeah it's right out front today," Ethan says looking at his phone. Most buses don't use bus stops anymore. Instead, this bus alters its pick-up locations based on the number of people who have requested a ride in the area.

A second chime on Ethan's phone echoes through the kitchen. "Bye, dad! See you tonight," he says as he runs out the front door.

You brush back the window curtain and watch as the school bus approaches. It eases to a graceful stop—perfectly parallel with the sidewalk thanks to the attentive algorithms of its navigation computer. Ethan hops on board, followed by the two twins that live next-door. Immediately, the doors close, and the bus soon departs.

This is the "free bus"—operated by the city. Its riders are limited to school children during the morning and afternoon hours of the week. But when class is in session, the bus allows anyone in the city to ride for free. The interior isn't lavish—consisting only of several drab rows of seats. There aren't any beverages, monitors, bathrooms, or audio speakers. And it doesn't vary it's route much—driving an endless loop around the city, picking up and dropping off passengers at approximately even intervals. Not very extravagant; but, on the other hand, it is free.

Ten minutes later, William, your oldest child, hobbles down the stairs.

"You missed the school bus," you say.

"I know. I can afford it," William says.

Missing the free bus means that he'll have to request a ride on an autonomous shuttle bus. The shuttles are smaller than the school buses—carrying about half as many people. They're not free for children, but it will only cost William a couple dollars to get to his high school.

The phone in William's hand chimes. "They're picking up at the corner," William says as he glances at his phone. "I better go. See you later!"

Leaning back in your chair, you brush back the curtain again and watch your second child hustle down the sidewalk. The shuttle slows to a stop near the intersection and opens its doors to let William inside. As it pulls away, you think back to the mornings of your youth—when missing the school bus incited a minor family emergency. Frenzied parents had to shuffle their schedules or ask neighboring soccer moms if they had any room in their carpools for an extra kid. But things are different now. Such morning dramas never happen anymore. Autonomous public transport could be requested as needed.

As you finish your breakfast, you notice that it's time to start thinking about your own morning commute. But you're not going to school, you're headed to the office park downtown.

As you glance at your phone, you see a cluster of ridesharing taxis that are available in your *executive's plan*—the monthly transportation service package that is free for every employee in your company. This particular plan isn't the best on the market. Guests aren't allowed to ride with you for free, and international airfare isn't included. But the seat compartments are comfortable and it's a nice perk for employees who need to commute to the downtown office branch.

The Taxi

You tap on your phone and request a taxi pick up in five minutes. Then, grabbing your notebook, you head out of the kitchen and into the game room—where you had stashed your briefcase last night. Until recently, the game room was actually a two-car garage. But the family hasn't owned any cars in years. The garage functioned as a storage room for a while. But now, the walls have been insulated, the floor has been carpeted, and two bay windows protrude out from the gap where the massive garage door used to stand. The renovated space wasn't originally meant to be used exclusively as a game room. But, as soon as the big screen was set up, that's what it became.

After rifling through a layer of blankets, soccer cleats, and discarded potato chip bags, you finally discover your briefcase. As you're placing your notebook inside the interior leather pocket, your cellphone chimes. A red message emanates from the screen reading:

We noticed that you're not standing at the pick-up location yet? Your car will be there in a minute. Would you like to reschedule?

You tap the "No" button on your phone and the message is replaced with a timer—counting down the number of seconds remaining until your car arrives. Your cellphone is always broadcasting your current location to the inbound taxi, and you're required to be near the pick-up location when it arrives so that the other riders don't need to wait.

Exiting through the game room's side door, you proceed to the pick-up point. The cars in your travel plan offer front-door service, so all you have to do is stand on the walkway in front of your house. A few seconds more and your taxi sneaks up on your left side. It eases into position along the curb—perfectly parallel just as your child's morning school bus was. But this vehicle is much smaller. It carries just four single riders—each in their own compartment labeled "A" through "D."

As the car comes to a stop, Door "C" sweeps open—revealing a cushioned seat and a row of illuminated lights—beckoning you to step inside.

You enter the vehicle and place your briefcase in a storage bin on the left side of the seat. A moment later, the door slides shut and the car's electric motor hums to life—thus starting your morning commute.

You tilt your head to the left so that you can scan the length of the cabin. Every seat in the car is occupied now. Each passenger is likely headed to offices in the same business park or to neighboring locations downtown. Because of the seating arrangement, social interaction with your fellow riders is not encouraged in this cabin type. But there are other types of ridesharing platforms:

- Some cars feature circular seating configurations for more gregarious riders.
- Some cars join seats together—encouraging couples or families to occupy an entire row.
- Some cars even pair riders up on a blind date—allowing them to chat each other up on their way to work. Later, the app provides an opportunity to exchange contact information if the encounter went well.

The commuter taxi you're in now has a standard linear seat configuration—commonly preferred by sleepy office workers who aren't too keen on socializing with strangers at the morning hour.

As the car accelerates, you push a button on the seat in front of you to release a slide-out table for your notebook. While going over a proposal for a new client, the car encounters a sudden change in inclination—indicating that you've just entered the freeway onramp. You glance through the window, remembering the days before self-driving cars—when you used to queue up at this very onramp, gripping the steering wheel of your Honda, staring at a smog-filled haze, and waiting for your turn to accelerate onto the freeway. Your commute to work used to be a 45-minute demolition derby through rush hour traffic—which seemed to get worse with each passing year.

But today, the road before you is teeming with autonomous taxis, shuttles, buses, semi-trailer trucks, and package couriers; there is no queue, there is no smog, and there is no traffic congestion. Each vehicle is harmoniously speeding down the road—guided by the watchful eye of A.I.

Your taxi speeds up to join the flow of traffic and squeezes in between a bus and a delivery truck. As each vehicle merges together with the mechanical efficiency of a zipper, you return your gaze to the document that's opened on your notebook.

Ten minutes pass and you hear the familiar hum of the taxi's electric motor as it accelerates once again—this time to its maximum cruising speed. The car is now entering a broad stretch of highway that connects the 'burbs to downtown. There are fewer turns and onramps on this segment of road. So, when the traffic is light and the weather is mild, the navigation computers allow some of the vehicles to move at higher top speeds. Most robo-taxis can achieve speeds of around 100 miles-per-hour under such conditions. You take a moment to glance out the window and acknowledge the flash of trees and foliage whizzing by at high velocity. But the scene is so unremarkable to you now that you soon lose interest and return to your work.

Another ten minutes later, and your car is approaching the downtown office park. A tone sounds in the compartment of the passenger in front of you—indicating that he should prepare to disembark. He will be the first drop-off this morning. The taxi coasts down an off-ramp and pulls up to an unfamiliar office complex with a bizarre sculpture in the entranceway. As you're contemplating the statue, your fellow rider collects his things. When the car comes to a stop, his door slides open with a whisper and he disembarks. The car soon senses the newly vacated compartment and the door slides closed. With the cabin secured again, the taxi proceeds down the road.

Your stop is next. As your office building appears on the horizon, you close up your notebook and retract the slide-out table. The exit notification chimes as your taxi comes to a stop at the drop-off zone near the building's front entrance. As your door slides open, a flash of blonde hair catches your eye. A smartly dressed woman is exiting from seat "A." You realize that you've met her before but you can't recall her name. It was something like Ava, or Olivia, or…

Suddenly another alarm sounds in the taxi, causing you to whip your head around in surprise.

"I think you forgot something," the blonde says as she walks by with a smile and a nod.

You throw a confused glance her way while simultaneously realizing that you exited the taxi with your notebook but not your briefcase—which is still sitting in the storage bin adjacent to the seat. The car's sensors are directed to scan for forgotten items each time a passenger vacates the compartment. If a misplaced article is detected, it beeps to notify the exiting party.

"Oh," you mumble as you make your way back to the taxi. Your mind hustles up an incident from your college days when you left your backpack on a bus that contained your wallet, keys, and about four hundred dollars' worth of textbooks—none of which you ever got back. But today, you retrieve your briefcase with a rush of relief—nicely averting a similar catastrophe.

Back at the curb, you crack open the briefcase to slip in your notebook. The robo-taxi's door slides shut and it sprints away to drop off its final passenger.

Strolling up the path to the building's main entrance, you glance over to see a demolition team knocking down the crop of concrete pillars that used to support the old employee parking structure. Given that you've been with this company for over a decade, you can readily recall the days when this interminable labyrinth marked the final challenge of your commute. After forty-five minutes in rush-hour traffic, you'd spend the next ten minutes spiraling up the towers in search of a parking spot. After finally locating one, you'd ride down the elevator in a frenzied state, cursing the day that your wife convinced you to buy a house that was so far inland.

But now, as you view the fleets of orderly robo-taxis queuing up to eject their cargo of well-scrubbed office workers, you take a moment to contrast the tranquility of your current mood with your old memories of morning molar-grinding stress—which would reverberate up your spine after the completion of each traffic jammed commute. The drive in used to be the *worst* part of the job. But, these days, it's a welcome respite from the daily grind.

As you approach the frosted glass door of your department, you can see the familiar profiles of your coworkers. The sound of the sliding door triggers your brain and a flurry of work-related queries rushes in to fill your stream of consciousness. As the automatic door slides shut behind you, its swift hush signals the end of your morning commute and the start of your workday.

The Ride Home

When five o'clock approaches, it's time to start thinking about heading home. Repeating the same process that got you to work this morning, you request a robo-taxi with your cellphone—this time selecting your home address as the destination. A confirmation message illuminates on the phone's screen—directing you to meet your taxi in front of the office building in six minutes.

"See you tomorrow," you holler at a coworker at a nearby desk. He gives a perfunctory wave as the door slides open and you exit out to the elevator lobby.

During the ride down to the taxi pick-up zone, your cellphone chimes again—indicating that your car will be approaching soon. This is the peak of evening rush hour so there are many taxis available in the downtown area—standing by in preparation for the mass exodus to the suburbs.

Departing from the lobby, you spot your taxi approaching the pick-up zone. This isn't the same vehicle that carried you to work this morning. But it's the same model and it's owned by the same fleet operator—so it's virtually indistinguishable. As the car comes to a halt, your phone chimes again—indicating that it's time to hop on board. The phones of two other people standing beside you chime as well. They don't work in your office, but you recognize their faces and nod as doors A, C, and D slide open. Compartment A will be yours for the ride home. The sight of its plush seat incites a sense of relief. You've made it through another workday and it's time to relax.

You enter the compartment and place your briefcase in the adjacent storage bin while making a mental note about the absentmindedness that caused you to forget your briefcase after the morning ride. As soon as your fellow riders are situated, the taxi doors close and the vehicle initiates its controlled acceleration.

You immediately push a button to dim the windows so that they are quite nearly black. The push of another button directs your seat to grind into a reclined position. As the car enters the freeway onramp, you put your headphones on and tap your cellphone screen to bring up a playlist of music. An acoustic guitar rift floats into your ears—prompting you to close your eyes and drift into a shallow sleep for the ride home.

Twenty-five minutes later, you rub your eyes and notice that the taxi is cruising down the familiar streets of your neighborhood. Your thoughts quickly turn to your children. Grabbing your phone, you pull up their vehicle activity log on your transportation app. Tapping on today's manifest, you can see their most recent trips. The younger child, Ethan, is currently at home. But he didn't take the bus. It looks like he took a shuttle to the food court (probably to hang out with his friends after school), and then he took another shuttle home. Your eldest child has scheduled a taxi pick-up for 6:30 p.m. That sounds like a long time to be at school, but you quickly recall that he has soccer practice on Tuesdays.

Soon, the car turns down your block and the destination notification alarm fills your little compartment. The tinted windows suddenly become transparent, letting in the light of dusk. And the rumble of spinning gears can be heard as the seat's backrest is returned to its upright orientation. Finally, the lights in your compartment illuminate and the car comes to a stop in front of your home.

Tuesday Evening

Grabbing your briefcase you step out of the taxi and admire the welcoming glow radiating from the array of lights that hover above your brick walkway. The home's computer has been notified of your proximity and it has adjusted the lighting for your arrival. As you step onto the porch, you can hear the abrasive sounds of machine-gun fire and explosions emanating from the game room. So, instead of entering through the front door, you take a detour and use the game room's side entrance.

Upon entering, you see your youngest child sitting on the couch—engaged in brutal online combat. As usual, he has the volume cranked up to eleven.

"Do you have to play that game so loud?" you holler, as you rush over to the console and crank the volume knob to the left.

"Oh, I didn't know it was so loud," he mumbles—barely acknowledging your presence.

You slide out of your coat and place your briefcase on a nearby table. Sitting down on the worn sofa next to your son, you put your feet up and stare at the gore on the video screen. As a parent, you feel it's your duty to offer the obligatory question, "Did you do your homework?"

"Starting it soon," Ethan mumbles back.

The answer is satisfactory enough for the moment. Feeling a twinge of hunger, your thoughts turn to dinner. You rise from the couch and turn toward the kitchen. "Did you already eat?" you ask.

"Yeah. We bought poke bowls at the food court after school," Ethan replies.

You enter the kitchen and the lights illuminate as you open the refrigerator to grab a Coke and consider your dinner options. A stack of prepared meals lies before you in individually sealed airtight packaging—ready to heat-and-eat. Healthy stuff. But at the moment, none of these options catch your fancy.

As you take a breath and ping your taste buds, you consider a memory of the fish and chips that they sell near the waterfront across town. You know it's not the healthiest choice. So you glance down at your diet app which auto-logs the price and nutritional content of every meal that you purchase. You can see that you've already eaten 1,700 calories today. The fish and chips plate of your desire has an unsurprisingly low health score. The app suggests a salmon dinner as an alternate meal. The compromise sounds reasonable so you alter your choice and order a plate of salmon and rice instead.

Exiting the kitchen and stepping into your bedroom, you peel off your slacks and hop into the bathroom for a quick shower. A few minutes later, you trade your towel for a conformable pair of sweatpants and return to the game room for some quality family time.

"Are you gonna let me play this time," you ask your son while sitting down on the couch near him.

"No, you don't know how to play this one," he says.

"Hey, I've been playing games like this for longer than you've been alive," you retort.

"Not this one. You…"

Just then your phone chimes—indicating that your salmon dinner is arriving.

"You've been saved by the bell for now," you say as you get up from the couch and head into the hallway. A vestibule extends out from your home's main entrance. It functions as both a porch and a dropbox for incoming home deliveries. The dropbox's port door butts up against the sidewalk so that it is easily accessible to autonomous delivery vehicles.

As you open your front door, you spot a courier vehicle pulling away from the curb. A green light shines above the vestibule's hatch—indicating that there is a package inside. You pull the handle and retrieve the box that holds your salmon dinner, as well as a smaller box that was probably dropped off earlier this afternoon—possibly containing the goalkeeper gloves that your eldest child had ordered for soccer practice.

You grab both boxes, walk back through the front door, and return to your seat on the couch in the game room. Opening the container's lid, a small burst of steam rises from the newly-cooked salmon.

"Are you sure you're not hungry?" you ask your child, feeling a twinge of parental responsibility.

"No, I'm ok," Ethan says while still engrossed in his game.

You grab the fork that came with the delivery box and put a slice of salmon in your mouth. Suddenly you're glad that you decided not to get the fish and chips.

The Product Purchase

"Dang! I got shot again," exclaims your child as his video game character is blown across the big screen.

"Well, why did that happen?" you ask.

"It's this controller," he says.

"You're not really going to blame it on the controller are you?"

"This one doesn't have the new crouch button that this game needs. Can I get the new one?"

"Your mom got you that one for Christmas," you reply.

"I know but she didn't get the right one."

"Well, how much can you sell this one for?"

"I looked it up last night. I can still get twenty-bucks for this one. Which means I'll only need another ten for the new one. So can I get it?"

You take a moment to consider your son's request. "If I pull up your class attendance log right now, am I going to see any late days?"

"I haven't been late for school a single time this quarter—unlike you-know-who! He's spent more than ten dollars on shuttles alone this month."

You smile with pride for a moment—realizing that his math likely does check out and his skills of negotiation are improving.

"Good point," you say. "Ok, you can get it."

"Yay," he says as he pauses the game and pulls out his phone. After a few moments of tapping, he looks up and says, "It's $31.24 ok?"

"Ok," you say.

With a final tap on his phone, the purchase is complete. "The new controller is on the way," Ethan says beaming. Just then, your own phone chimes—indicating that a sale has just occurred via the family account.

"Time to get rid of this old one," Ethan says as he flips over the game controller in his hand. He points his phone at a metallic strip that lies along the bottom of the controller. Immediately, a confirmation tone emanates

22

from his phone—indicating that the product has been successfully entered into the online marketplace.

"Gotta get a box," Ethan says as he leaps off the couch. He runs to the closet, flings open the door, and grabs a small shipping box from a stack of dozens resting on a shelf. He places the game controller inside the box and secures the adhesive lid that lines the top. He then waves his phone over the box's address identifier and a grey checkmark appears on the label—signaling that the box is now ready to ship.

"How much money did you get for the old game controller?" you ask.

"$19.27," Ethan says.

"Close enough," you say with a shrug. "Where's the buyer located?"

"I don't know. Someplace in Arizona, I think," Ethan says as he steps out through the front door and into the vestibule. Once there, he drops the box into the vestibule's hatch and presses a button indicating that a pick-up is needed. This is not a time-sensitive shipment. So the mail courier vehicle probably won't come around until much later tonight—when most people are sleeping and the roads are empty.

As Ethan turns to enter the house, you offer a suggestion. "While you're waiting for your new toy to arrive, why don't you put some food in your mouth so that I can tell your mom that I watched you eat dinner when she comes home?"

"OK," he says in compliance, skipping off to the kitchen.

As you return to your own salmon dinner, you think back to a time when selling previously-used items on the internet was a chore that might take a week to complete—often involving complicated product listings, costly delivery fees, and waiting in line at the post office. But now that autonomous couriers move everything so swiftly from city to city, even your youngest child is able to list and sell products on the secondary market with ease.

As you're shoveling the last spoonful of rice into your mouth, your phone chimes. In the distance, you can faintly hear an identical chime on Ethan's phone.

"It's here!" he shouts as he runs from the kitchen, through the front entranceway, and out to the vestibule again. Seconds later, he reemerges with the box containing his new game controller. He rips it open, reaches inside, and pulls out the sleek new device with an admiring look.

"Happy now?" you ask.

"Yah! Thanks, dad," he says as he taps a few buttons and links the controller to the game console. A moment later, the screen comes to life and Ethan resumes his game.

"Now are you going to let me play?" you ask.

"Sure! As soon as I'm done," he says with a grin.

Let's Review

In our preceding futurama, we described a day in the life of a typical middle-class family—the beneficiaries of a fully autonomous infrastructure. Several technological conveniences are described in the narrative; we'll be discussing each one throughout this book. But for now, let's briefly call out five noteworthy innovations.

#1: The family used autonomous transport to get around

Each of our three family members took a different type of vehicle to work or school.

- Ethan (the punctual child) used the city's free autonomous school bus.
- The eldest child took an autonomous shuttle to school.
- And the father took an autonomous ridesharing taxi (a "robo-taxi") to work.

Even though one child missed his morning school bus, the incident didn't result in any logistical dramatics. Instead, the late child simply pulled out his cellphone and requested an alternate pick-up. Take note of the modest transportation fees in our story. The second child paid a few dollars for an autonomous shuttle, the father's robo-taxi was included as a company perk, and the youngest child's bus ride was free. Regardless of the type of payment schemas that autonomous platforms will have on offer in the future, the price is sure to be lower than the current transportation costs of the average American household—which (in 2017) The Bureau of Labor Statistics estimated to be $9,576 annually. The ability to request a ride "as needed" and to "pay-as-you-go" will render *personal car ownership* uneconomical in the decades to come.

#2: The family didn't need a garage

Because the parents no longer owned any cars they no longer had any use for the family garage. Recall that the father in our story had converted the family garage into a game room. A typical two-car garage (24' x 24') takes up 576 square feet in an American home. When robo-taxis become the preferred mode of transportation, homeowners will no longer need to devote all of this space to vehicle storage. Unlike conventional autos, robo-taxis don't need to be parked. After dropping off a passenger they get back to work—returning to the road to retrieve a new fare.

The many inefficiencies of car ownership are made evident when you consider that the US is home to over 272 million registered vehicles, and that all of these vehicles will spend 96% of their lives sitting in a garage or parking space. Given the amount of time, raw materials, technology, and labor that goes into the construction of these machines, it is a travesty that so much of their potential is squandered—mostly devoted to providing shade for asphalt parking spaces or concrete garage floors.

In the coming autonomous age, we'll be able to eliminate this waste. We'll be able to do away with parked cars, parking lots, and home garages all together.

#3: No traffic congestion or accidents

Because the father in our story used a robo-taxi to get to the office, he didn't need to expend any energy driving through morning rush-hour traffic. Instead, he was able to pull out his notebook and get some work done during the drive. Note that the average American spends 19 workdays per year sitting in traffic. Consider what could be accomplished if we were to give this block of time back to every citizen in the country.

Most importantly, during the father's morning commute, he didn't encounter any traffic congestion nor any traffic accidents. Since Henry Ford's Model T left the factory floor on August 12, 1908, over 3.7 million Americans have been killed in auto collisions. The National Highway Traffic Safety Administration estimates that 94% of accidents are the result of human error—the three leading causes being alcohol, speeding, and distraction.

Given the nature of driving and the limitations of the human attention span, it is irrational to expect people to drive responsibly during every single moment behind the wheel. Even the best drivers make mistakes or succumb to the sheer tedium of uneventful steering. Of all the arguments for self-driving cars, the yearly auto fatality statistics (of around 36,000 Americans per year) are often the most persuasive. Let us hope for a world in which the cost of transportation does not demand the sacrifice of so many lives.

#4: Rapid product delivery

In our story, the youngest child was able to purchase a game controller online and have it delivered in minutes. It's no secret that large online retailers (like Amazon and Walmart) currently have their eyes glued to the horizon—diligently scanning for the headlights of the coming fleet of autonomous delivery vehicles. When such stores are finally able to ditch their reliance on human-driven delivery trucks then customer orders will be fulfilled at lightning speeds. Boxes will be passed from one warehouse robot to the next, and then loaded onto autonomous couriers—who will then race your brand new game controller (or beard timer, or towel rack, or cat box) to your front porch. Someday, the entire order-fulfillment process will not require the participation of a single human hand.

#5: Hot meal delivery

After arriving home for the day, the father in our story purchased a salmon dinner—which was promptly delivered by an autonomous courier. When it comes to take-out food, our civilization has a bizarre knack for delivering pizza. But for every other food product, efficient home delivery remains an elusive goal.

Driving a hot meal across town is trickier than you might think. But, when the coming autonomous infrastructure is complete, the majority of *restaurant meals* may actually be consumed at home. Every eatery will have an online menu and autonomous couriers will queue up to rush each readymade meal to the customer's porch.

Not all meals will be prepared individually; most will be mass-produced in large quantities—thus reducing the per-unit cost. Someday, the delivery of prepared meals may be so efficient and economical, that future houses will have little need for a kitchen.

When considering the future world of autonomous vehicles, we tend to focus on their utility as "people movers." However, they will probably be even more useful for the transportation of objects—mostly food and retail products. Our future will indeed feature millions of self-driving passenger cars—carrying people to work or school. But there will be even more vehicles devoted to carrying parcels—dropping off your mail, your groceries, your latest purchase from Amazon, and the meatball sandwich you ordered from Togo's five minutes ago.

Unlike King Solomon—who needed a flying carpet to get to his favorite restaurant in Damascus—the transportation marvels of *our* future will be carrying the restaurant meal to us.

Suburbia of the Future

The trick to understanding your *future life* lies in envisioning how your *current life* would be different if the chore of mere movement was not such a limiting factor—i.e. if the act of just "showing up" was not such a costly spectacle.

- If your car "drove itself" then what would you do on your way to work?
- If shipping costs were free and all deliveries were fulfilled in thirty minutes or less, then how would this change your buying behavior?
- What would your house look like without a garage?
- How would people and parcels enter your home if you didn't have a driveway?
- What would your neighborhood look like if there were no parked cars on the street?
- Would you be more likely to visit someone you know if you didn't have to drive so far to see them?

The following image depicts what an idealized future neighborhood might look like after the adoption of autonomous vehicles and their supporting infrastructure.

What do we see in this image?

- Instead of several crisscrossing roads, there is just one small road—located between the homes. There are no parked cars or garages in our neighborhood. Instead, the vestibules of the houses meet the sidewalk—thus enabling autonomous couriers to deposit their deliveries from the road.
- Speedbumps and ancillary parking lots have been removed—robo-taxis won't be speeding nor parking in the future.
- Pedestrian paths, bike roads, and trails weave through the neighborhood—unhindered by crosswalks and stop signs.
- The broad avenues and intersections have been replaced with open grounds—converted into a soccer field or left to return to a natural wooded state.

We don't know how future city planners will manage to retrofit the appliances of the autonomous infrastructure onto the sidewalks of our existing neighborhoods. But it is clear that, in order to fully capitalize upon the efficiencies of the autonomous age, engineers will need to rethink the way they build roads, houses, and cities. When these machines finally arrive, they will integrate into every aspect of our lives and change the way we think about movement, distance, and commerce.

By now, I hope it is clear that self-driving cars are much more than just a robot chauffeur screwed to a driver's seat. Instead, self-driving software will be loaded into millions of wheeled machines—of multitudinous shapes, sizes, and configurations. They will dance across the streets of your future neighborhood, wondrously dropping off gifts at each doorstep—with a knack for home delivery that rivals the skillset of Santa Clause.

If we can manage the transition to an autonomous economy and if we can successfully modify our cities to utilize the benefits of this emerging technology, then a cleaner, more spacious, more efficient, more charitable, and, yes, a more beautiful world awaits. However, we should note that there are many potholes on the road to Autopia.

The Great Disruption

The copacetic narrative that begins this chapter only portrays a *possible future* for America—one of many potential outcomes. We stand now on the cusp of a great transportation revolution. Successfully managing civilization's metamorphosis—to a society in which man and autonomous machine peacefully coexist—will be a delicate affair. And unfortunately, for many people, things will get worse before they get better.

Figure 6 - Volvo's autonomous semi-truck prototype—code named: "Vera."

Many a breathless column has been written by tech journalists who dutifully describe how self-driving A.I. is about to steal every trucking job in the country—which (according to the U.S. Census Bureau) accounts for 3.5 million jobs. Such fears are not misplaced. All of our nation's freight will be hauled by machines someday. These jobs will eventually be pinched. But this is just one of many blows to come. Thousands of industries exist solely to facilitate the movement of parcels and people down the road. Consider the taxi driver (Uber and Lyft), the delivery driver (UPS, FedEx, DHL), the bus driver, the DoorDash driver, and the mailman. These jobs will be in jeopardy as well.

Most importantly, consider the massive infrastructure that exists to support all of these vehicles. The truck transport industry alone employs two million mechanics, managers, and dispatchers. But this number pales in comparison to the number of industries that cater to you and me—the personal car owners. Consider the auto mechanic, the auto dealership, the auto supply store, the car wash, the garage door installer, and the gas station. When robo-taxis become the primary means of transportation, none of these industries will exist anymore—at least not as we currently know them. And now, stop to consider all of the ancillary jobs that these businesses rely on:

- The accountant who does the books for the car wash.
- The line worker who assembles garage door openers.
- The janitor who cleans the bathrooms at the gas station.

These people will lose their jobs too.

Pundits have been quick to recognize the threat that autonomous technology will impose upon truckers. They are correct to do so; many men have put shoes on their kids' feet with money that was earned spinning a wheel. But most commentators have neglected to infer the next step in this disruption. When self-driving A.I. is finally perfected and when pneumatic drills begin ripping steering wheels out of big rigs, the screams of their gears will reverberate throughout every job sector in the country—affecting every worker, whether he's employed in the transportation sector or not. Much of the US economy is supported by dominos that stand atop the vast corridors that connect the nation. Jolting these corridors will inevitably cause many proximal dominos to topple over. And nobody knows when the rally will end.

Stages of Adoption

While there may be long patches of darkness on the road ahead, the fruits on offer at our final destination could be well worth the trip. It is possible that autonomous technology will help to usher in a time of great abundance—wherein mankind will become so adept at producing and delivering life's necessities that their price can be whittled down to a mere nominal fee.

The wonders of the autonomous age will not fall into our lap overnight. The rate of their adoption will be determined by a confluence of economic, political, legal, and cultural forces. During this process, the nation may enter a time of great polarization—in which some municipalities welcome self-driving cars with open arms, while others consider them to be an abomination and a curse.

- In some regions, autonomous vehicles may not even be allowed to cross the county line.

- In other areas, the citizens may not allow human-driven cars onto their streets.

- Finally, in some locales, a third option will be tried in which man and machine are expected to share the highways in harmony.

Ironically, this third strategy may turn out to be the most calamitous. The morning commute in such towns may become a precarious game of "man versus machine"—in which human drivers grow increasingly intolerant of any driving missteps committed by their robot competitors. It's easy to imagine such contests going awry—resulting in roadways that are *more* dangerous and *more* congested than ever before.

Aside from such logistical objections, many of our countrymen will not only perceive autonomous vehicles to be dangerous or disruptive, but also as just downright un-American. Our national identity has evolved around the purchasing and showcasing of prized automobiles. That shiny red sports car in the driveway has long been the symbol of American ingenuity, prosperity, and freedom. But when everyone gets around town on ridesharing robo-taxis, automobiles will become just another appliance—capable of inciting about as much emotion as a blender or a microwave.

Some Americans aren't too happy about that. They will never relinquish their car keys and they are already gearing up to protest against the coming onslaught of four-wheeled robots. On the other hand, many are sick of sitting in traffic and breathing in smog. They are looking forward to the day when their family sedan can be replaced with an autonomous electric robo-taxi. The more environmentally-minded of the lot may soon be campaigning to make internal combustion engines illegal altogether.

Perhaps you don't know which team to route for yet. Most of us don't. We still have some time to decide. But regardless, let us not doubt that autonomous vehicles are coming to a town near you. And they are about to change our lives in many ways.

- They will change the way we live, work, and play.
- They'll change the way we shop, cook, and eat out.
- They'll change the way we display our wealth and status.
- They'll change the way we raise our children, the way we choose our neighborhoods, and the way we navigate the world with our spouse.
- They'll change the way we think about work, charity, and poverty.
- And, most importantly, they'll change the way we think about each other.

Many of these changes will be good and welcomed. Others will be scary, and they will force us to rethink some ancient existential questions—about the nature of community, about what defines a relationship, the proper role of human labor, and about the very purpose of our existence.

Autonomous vehicles will be the most disruptive technology to ever be unleashed upon humanity.

This book is about that disruption.

Let us imagine…

For now, let us dare to dream of a better world.

In the first half of this book, we will be focusing on the innovations that will be available to us after the autonomous infrastructure is finally completed. At this stage, every car will be built without a steering wheel and every house will be built without a garage.

In the latter half of this book, we will try to identify some of the many hazards that lie on the road to the autonomous city of tomorrow. And, we'll suggest how we might circumnavigate some of the more challenging quagmires.

Any jaunt through the realms of speculative technology is bound to result in bouts of prognostication that turn out to be laughably inaccurate. But let's take a ride anyway.

Let us imagine a world of self-driving cars. One in which commuters can nap on their way to work—gliding down a road that is free of congestion and auto accidents. Let's imagine a busy intersection with no traffic signals—through which cars jeté in a balletic display of computer-controlled choreography.

Let us imagine a world in which the litter of parked cars has been cleared from every curbside—hence revealing the grand avenues of our cities for the first time in a century. Imagine suburbia without garage doors—those drab planks that stand as the unfortunate architectural pièce de résistance of the American home.

Let us imagine a world where our mailboxes have been transformed into intelligent delivery appliances—capable of simultaneously shipping outgoing parcels and receiving incoming pizza deliveries.

Imagine a world with more elbow room. A world where long-distance transportation is so efficient, that we no longer need to cram people together into congested spaces—solely for the purpose of housing them near their place of employment.

Imagine a world where almost anyone can afford to buy a home. A world in which it is no longer impractical to purchase a parcel that is *far* outside the city center—where the land is less expensive, and the horizons are broader.

Let us imagine a world where the homeless need not sleep on the sidewalk—merely to remain in proximity to the point at which charity is dispensed. Instead, imagine a world where charitable offerings drive themselves to the needy, wherever they reside.

Let us imagine a world in which every engine is electric and every skyline is free of smog.

Let us imagine a world in which access to rapid transportation is not a privilege, but a right.

Imagine a world in which *every* vehicle is an autonomous vehicle.

Let us imagine…

Ch. 1: Getting Around Town in the Future

America was home to a few thousand horseless carriages and motor cars at the dawn of the 20th century. But it wasn't until Henry Ford's Model T rolled off the assembly line in 1908 that the automobile would forever change the world of transportation. Over the next decade, Ford would manufacture 15 million cars—a record that stood until 1972 when it was finally overtaken by the Volkswagen Beetle.

In 2019, the Department of Transportation estimated that the U.S. had 225 million licensed drivers operating 272 million registered vehicles. Worldwide, the number of cars on the road is somewhere around 1.4 billion.

Obviously, these four-wheeled machines have thoroughly integrated themselves into our economy and into our lives. Cars and trucks have come a long way since the Model T. The features currently available in the lowliest junker are better than anything Henry Ford could have imagined in his time. But where do we go from here?

What is the next stage in the evolution of the automobile?

If you were to try to predict the future of transport by watching Hollywood films, you'd conclude that our skies are soon to be filled with "flying cars"—gravity-defying vehicles that are capable of casually intermingling between airborne traffic and skyscrapers. But flying cars will *not* demark the next leap forward for humanity—autonomous vehicles will.

Figure 7 - Flying police cars and taxis are depicted in the 1997 film "The 5th Element."

When most people think of self-driving cars they imagine their own car—but with a button that enables it to *drive itself*. But this doesn't adequately describe the many types of vehicles that will be featured in this book. Instead, when we refer to *self-driving cars* or *autonomous vehicles*, we're referring to machines that are capable of "Level 4 Autonomy." According to the Society of Automotive Engineers (SAE), cars at this stage can be built without a steering wheel—thus they require absolutely no navigational input from the driver. We'll take a moment to discuss all six levels of autonomy now.

The Six Levels of Autonomy

Level 0 autonomy describes a vehicle with no self-driving capabilities at all. However, the car may include some intelligent features—many of which you might currently have in your own car—like blind-spot detection or emergency braking.

Level 1 cars can instigate small steering or acceleration inputs—enabling the car to do *lane-centering* or *adaptive cruise control.*

Level 2 cars are capable of a modest degree of autonomous highway driving, but they still require constant road-scanning by the driver. This degree of autonomy has been available in Tesla's "Autopilot" feature since 2014. With this technology, drivers may release the steering wheel on some roads, letting the A.I. take over for part of the journey—often on stretches of uneventful highway. Because Level 2 cars are so good at creating the *illusion* of total control, many consider them to be too dangerous for the road. Readers may have seen one of many news reports featuring photographs of Tesla owners asleep at the wheel.

Level 3 cars can function autonomously and the driver is permitted to divert his attention away from the driving task—he can watch a movie or read a book while en route. But the driver must still be prepared to take the controls if called upon by the onboard computer. Like Level 2 cars, Level 3 cars can operate autonomously for the majority of a highway road trip. Hence, they tend to incite driver complacency. Many experts believe that Level 2 and Level 3 cars s hould never be mass-produced—and that consumers should just wait for vehicles that are capable of Level 4 autonomy.

Level 4 cars are fully autonomous. They do not require any input from the driver—so they may be constructed without a steering wheel and pedals. Additionally, the driver is not required to intervene in case of an emergency; he can even sleep during the trip if he likes. However, car operation may be restricted to designated locales and certain road types. When the conditions for autonomous travel are not appropriate, Level 4 cars must be capable of aborting the trip—i.e. they must park themselves and stay put until the appropriate driving criteria are met.

Level 5 cars can operate on any road and in any weather condition that a human driver could be expected to negotiate. The driving skills in a Level 5 car should be as good (or better) than those of a professional human driver.

Cars without steering wheels

Since early 2019, Teslas have come equipped with all of the hardware needed for fully autonomous driving. But their software is still being perfected. For now, all Tesla autos still require a steering wheel and pedals. Except for their large dash-mounted touch screens, Teslas look very much like conventional autos. But someday this will change.

In the future, the majority of cars will not include a steering wheel. Eventually, human driving inputs will not even be possible. To get an idea of what the interior of such cars might look like, consider the two-seat prototype vehicle that came out of Google's "Firefly" project. In the following diagram from Google's 2016 design patent, note that the steering wheel and instrument panel have been eliminated and replaced with a baggage storage bin.

Patent No.: US D770,349 S
Date of Patent: Nov. 1, 2016

Such designs are representative of the types of midrange autonomous vehicles that will be carrying you around in the future. They'll be capable of taking a small group of people on a short jaunt—like a commute to work or a drive into town.

Currently, such cars are not available for purchase—building a car without a steering wheel is not exactly legal yet. But Waymo has petitioned the National Highway Traffic Safety Administration to consider removing such regulatory barriers—so that they can at least test their cars (free of instrumentation) on public roads. When such regulations are lifted and when autonomous driving software is perfected, then such "hands-free" vehicles will eventually be the only type of passenger car on the road. Someday, riding in a car with a steering wheel will be as quaint as riding in a car with a hand-crank starter engine.

Why traditional carpooling sucks

If you've ever sat near a highway and watched the ebb and flow of commuter traffic, then you might make three observations.

- First, you'll notice that many of the cars are traveling in the same direction and headed to the same general vicinity.

- Second, you'll notice that the majority of passenger cars are capable of carrying five people, and yet most of them are occupied with just one.

- Third, you'll notice that the road predictably overflows with cars at about the same time each day—during "rush hour traffic" between the hours of 6 to 10 a.m. and 4 to 8 p.m.

Such travel patterns are particularly prevalent in US metro areas—where millions of people all get into their cars at around the same time, commute to work in the same general direction, and park in the same cluster of downtown lots. According to the U.S. Census Bureau's 2014 estimate, 76.4% of Americans get to work by "driving alone in a private vehicle." Meaning that billions of transportation dollars are solely devoted to hauling one person (and four empty car seats) up and down the highway—day after day.

In considering the above observations, politicians, economists, city planners, and environmentalists have long insisted that the solution to our traffic quagmire is crystal clear: *People just need to start carpooling more!* After all, given that there is almost always *someone* headed in the same general direction as you, then wouldn't it make sense to devise some sort of method for you to share your car with this person?

If we can squeeze multiple people into just one car, then there won't be so many cars on the road.

- That would mean less traffic congestion and less smog.

- The cost of operating the car could be distributed among each of the riders.

- Such a schema would be better for the environment and better for our pocketbooks.

- Carpooling is the intelligent resource management solution that we've all been waiting for!

Right?

Well, that's how it's supposed to work. But as we all know, that's never how carpooling actually works.

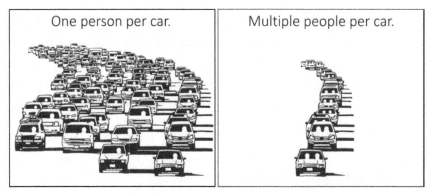

Figure 8 - Many people on the highway are headed to proximate destinations and most cars have five seats. Yet, because of the social and logistical complexities of carpooling, the vast majority of cars only carry one person.

The US Department of Energy has been tracking the decline of carpooling for decades. In 1980, 19.7% of Americans carpooled to work each morning. But by 2014, that number had fallen to 9.6%. Vehicle navigation and communication technology have substantially improved over the last quarter-century. And yet, people hate carpooling more than ever. This is primarily because the difficulties in carpooling are not due to any technical inefficiencies of the cars themselves. Instead, carpooling is hard because of the behavioral inefficiencies of humans.

Anyone who has had the pleasure of participating in a morning carpool already understands this quandary.

- Someone in the carpool is always running five minutes late.

- Someone can't drive on Tuesday, but they can drive on Friday.
- Someone forgot that they were supposed to pick up last week.
- Someone forgot to mention that they were going on vacation.
- Someone likes to eat in the car.
- Someone forgot their purse in the car.
- Someone is always too cold, while someone else is always too hot.
- Someone likes to sleep during the ride, while someone else is very chatty in the morning.
- And at least one person in the car invariably hates your taste in music.

Such are the joys of sharing a cramped space with your fellow man. In taking a moment to reflect upon the potential social and logistical issues associated with carpooling, it's no wonder that so many cars on the highway contain only one person. Most of us would prefer to fight traffic alone (in the tranquility of our own temperature-controlled cabin), than combat it with an infantilized squad of besuited coworkers.

Ridesharing Software

In trying to combat the preceding list of carpooling complexities, many solutions have been proposed. Ridesharing schemas actually originated over a hundred years ago—when (in 1914) L.P. Draper picked up a passenger in his Ford Model T in exchange for a nickel. At that time, nickels were sometimes called "jitneys"—whence a new industry was christened and a new type of travel swept the nation. At their peak, 62,000 Jitneys operated in 175 cities across the country.

Figure 9 - The first Jitney ride purportedly took place on the streets of Los Angeles in July of 1914. (Source: George Grantham Bain collection, Library of Congress.)

Competitive with bus and streetcar fares, but cheaper than a taxi, Jitneys could carry multiple riders and would occasionally veer off their regular route to offer front door service to passengers. Unfortunately, due to waves of regulation (purportedly enacted by cities that relied on tax revenue from streetcar companies), the Jitneys were mostly extinct by 1920. Though some variations still operate in Atlanta and Miami.

Now let's jump one century forward—to August of 2014. Uber has just introduced a new service called "UberPool." It's an ambitious undertaking that attempts to pair up passengers who happen to be headed in the same approximate direction. Shortly after its initial launch, Uber added "Uber Express Pool"—an even cheaper version of UberPool—wherein the passenger's pick-up and drop-off location may be a "short walk" from his actual origin and destination address.

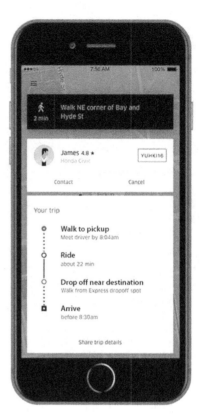

Figure 10 - To save time, *Uber Express Pool* **prompts riders to walk to an impromptu pick-up spot to meet their Uber driver.**

In describing this technology, Uber product manager Ethan Stock told The Verge:

> **We think carpooling is very much the way of the future… [We] think the transformation of car ownership towards carpooling is going to be tremendously beneficial for cities [and] for the environment… [Because of] all the reasons that we're [all] very familiar with—congestion, pollution, etc.**

In theory, UberPool is a great idea. After all, taxis have five seats. So why not fill the empty seats with other passengers who are headed in the same direction?

However, while such ridesharing schemas seem reasonable, complaints about UberPool are easy to find online. Uber drivers are quick to cite the many difficulties that they encounter when tasked with picking up multiple strangers. Along with all of the carpooling difficulties that we discussed in the previous section, the entire enterprise is plagued with additional logistical complexities:

- Delays are very common among taxi passengers. But with UberPool, the irritation of the situation is easily compounded. Because, if the passenger from the first pick-up location is running late, this bout of tardiness only irks the Uber driver himself. But when the *second* passenger is running late, his belatedness inconveniences *both* the Uber driver *and* the first passenger—thus inciting a socially awkward situation.

- Passenger tardiness isn't necessarily the result of thoughtlessness. The requesting passenger might be slow in arriving at the pick-up point because of a physical disability, or a luggage problem, or he simply might not be familiar with the area. Sometimes the place where the Uber app tells the requesting rider to wait is actually *not* a viable or safe pick-up location.

- And even if all goes well, and both passengers manage to settle into their seats on time, they often don't particularly like each other.

Such observations have led many Uber drivers to limit their fares with UberPool—instead preferring to carry a single passenger rather than undergo the stress of managing the demands of multiple riders.

Still, apps like UberPool are a good start. And they offer us a glimpse into the primary method by which robo-taxi rides will be hailed in the future. However, truly efficient ridesharing will only be available after the software is integrated with taxi vehicles that are specifically designed for passenger transport. In the next section, we'll describe such hardware.

Ridesharing Hardware

The successes and failures of the UberPool app provide us with a glimpse into the future of ridesharing. Traditional carpooling (squeezing four passengers into a *conventional* automobile piloted by a human driver) will forever be a tedious venture. But, thanks to self-driving A.I., future autonomous robo-taxi platforms should eliminate most of these shortcomings.

To capitalize upon the efficiencies of autonomous travel, ridesharing platforms will need both a software *and* a hardware upgrade. To transport multiple passengers in the most convenient manner possible, we'll need to redesign the taxi. In the below diagram, we have contrasted the seat configuration of a conventional human-driven taxi with a potential future robo-taxi platform.

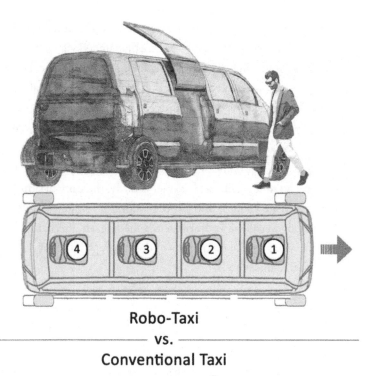

Robo-Taxi
vs.
Conventional Taxi

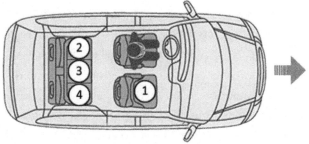

Figure 11 - Future robo-taxi designs will eliminate many of the problems commonly associated with ridesharing or carpooling.

Our autonomous taxi has a platform configuration that is designed to cater to individual commuters. (This vehicle is similar to the taxi hailed by the father in the narrative that began this book.) Here are some key features to take note of:

- First, notice that our robo-taxi does not contain a human driver. So, of course, the car does not require a driver's seat, a steering wheel, nor any vehicle instrumentation at all. All communication with the taxi is done via an app on the passenger's cellphone. Rides are hailed, routes are computed, and taxis are dispatched via the fleet's software. The system will be continuously taking in new traffic data, managing incoming passenger requests, and computing the ideal route needed to deliver each rider to their destination as needed.

- Unlike conventional carpools—who always find themselves waiting for that one person who is "running a bit late," the robo-taxi's computer will be notified if the rider is not standing at the pick-up zone on time. When a customer is tardy, the car will simply disregard his pick-up request and proceed along its original route. The tardy passenger will be prompted to schedule an alternate taxi. Because robo-taxis will be ubiquitous in the future, additional rides should be available in seconds.

- Since cars almost always drop off and pick up passengers on the right-hand side, robo-taxis have no use for doors on the left. Instead, each seat has its own door—which will open independently whenever a rider enters or exits the vehicle.

- Four passenger seats are visible in our sketch, though other designs call for more. The number of available seats in this vehicle is kept low so that riders won't have to tolerate an excessive number of drop-offs before reaching their final destination.

- Instead of placing seats in a horizontal formation (i.e. instead of forcing strangers to sit together), each passenger gets their own compartment. Thus eliminating the social workload associated with traditional carpooling. Similar to an airplane seat, each one of these compartments will be equipped with a fold-out tray or desk, personal ventilation, lighting controls, and a reclining seat—allowing passengers to sleep while they ride.

In considering the preceding list, such amenities are sure to make the chore of "8:00 a.m. carpooling" more palatable.

Why we will all use robo-taxis in the future

So long as there is a human at the wheel, carpooling will never be a viable commuting option for most of us. It will only be via the synergy of autonomous vehicles *and* ridesharing software that the efficiencies of autonomous travel will be fully actualized. Without this union, it is unlikely that apps like UberPool will ever be able to consistently provide a friction-free ridesharing service. But such benefits were probably only the pretense to justify the expense of UberPool's development. In constructing such apps, companies like Uber are looking to the horizon—waiting for the dawn of intelligent self-driving A.I. and the enactment of legislation that will allow autonomous vehicles onto public roads. When this day comes, ride-hailing companies (like Uber and Lyft) will be able to fire all of their human drivers and replace each one with an autonomous robo-taxi. This is not mere conjecture. Back in 2014, Uber CEO Travis Kalanick was forthright about his future plans:

The reason Uber could be expensive is because you're not just paying for the car—you're [also] paying for the [driver] in the car. When there's no [driver] in the car the cost of taking an Uber [will be] cheaper than owning a vehicle… And then car ownership goes away.

Instead of jumping from the shower, to the kitchen, to the family car (to start the race to work each morning), our future commutes will begin with our cellphones—which we'll use to request a robo-taxi pick-up. We'll all have to learn a new set of travel behaviors. Altering our morning rituals will require a bit of practice. And it will take years for Uber to teach their software to juggle multiple rider requests. This is why Uber has started the development process now—so that we'll all be ready when the age of autonomy is finally upon us. As Motherboard editor-in-chief Jason Koebler observed:

Right now, UberPool [is] not hyper-efficient because of the human element… But all the interpersonal problems…go away if the car has no driver. Uber [is] training us so we're used to these services when they get rid of drivers altogether.

The reason that Uber is so willing to take on this "training expense" becomes clear if you do the math. Personal car ownership is very pricey. In 2019, the

American Automobile Association (AAA) calculated that Americans are paying $9,282 annually to own and operate a medium-sized sedan. This is the highest estimate ever recorded since AAA started tracking vehicle expenses in 1950. Note that this sum doesn't include additional fees for driver training, traffic tickets, and parking violations. Nor does it include the cost of attaching a 576 square foot two-car garage to the side of your home. And remember, we're paying all this money for a device that spends 96% of its life sitting in a parking space and doing absolutely nothing. This is why the economics of autonomous ridesharing will eventually lure people away from personal car ownership. Instead of sitting in a parking lot, robo-taxis could be actively hauling passengers around all day—stopping only momentarily for a battery charge and a car wash.

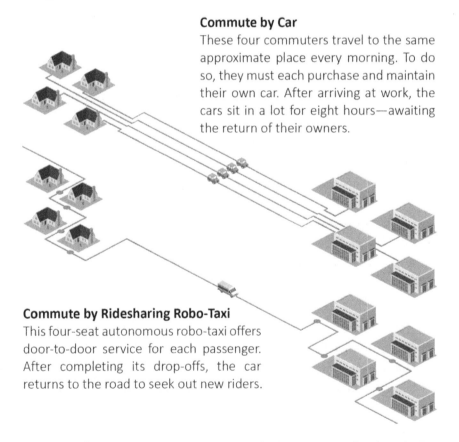

Commute by Car

These four commuters travel to the same approximate place every morning. To do so, they must each purchase and maintain their own car. After arriving at work, the cars sit in a lot for eight hours—awaiting the return of their owners.

Commute by Ridesharing Robo-Taxi

This four-seat autonomous robo-taxi offers door-to-door service for each passenger. After completing its drop-offs, the car returns to the road to seek out new riders.

Additional financial savings can be expected when we consider the role that dedicated robo-taxi do *not* have to play. Unlike conventional autos, robo-taxis need not be crafted to act as mechanisms for displaying wealth and status. Their gears won't be shifted via hand-polished mahogany knobs and their seats won't be adorned in Chrysler's *Corinthian Leather* (which actually comes from New Jersey). Instead, most robo-taxis will be mass-produced for efficiency and ease of maintenance. When individual components fail, they'll simply be exchanged for commoditized parts and the vehicles will be promptly returned to the road—in shipshape and Bristol fashion.

It will be difficult for future commuters to rationalize the expense of personal car ownership. The financial incentive—to exchange their car keys for a ridesharing app—will eventually be too great. Someday, the majority of passenger trips will occur in robo-taxis and the majority of Americans will no longer own a car. Instead, all of their rides will be requested and paid for via apps—many of which may look quite similar to the existing UberPool interface. This is why so many companies are ravenously investing in autonomous passenger vehicle technology. Because whoever controls the robo-taxi fleets of the future, will also control the transportation dollars of the 143 million Americans who commute to work and back each day.

Ch. 2: Why Would You Want to Ride in a Self-driving Car?

Most of us have sat through at least one late-night news broadcast heralding the eventual arrival of driverless cars. Many of us are looking forward to the day when the computer in our dashboard is smart enough to *take the wheel*— so that we can *take a nap* on the way to work. But the *self-drive* feature is just one of several that autonomous vehicles will have on offer. There are many ancillary benefits to be had in the age of autonomy—most of these are not readily evident at first glance. So it can be difficult to comprehend why some researchers believe that self-driving technology is about to "change the world" (as the title of this book reads). In this chapter, we'll list ten reasons that describe why we believe this to be the case; why we believe that, someday, *every* vehicle on the road will be an *autonomous vehicle*.

Reason 1: Self-driving cars do not require cognitive effort to drive

The primary benefit of self-driving cars is that they can drive themselves.

Of course.

They don't require any input from a human driver and hence, no mental effort is expended during the commute. Driving is a cognitively demanding task. Even professional truck drivers are only allowed to drive for 11 hours per day. When we wake each morning, our mind is presented with a finite reserve of brainpower—which (ideally) can be apportioned to productive pursuits. Unfortunately, much of this energy is squandered during the morning rush to work, and further depleted during the evening drive home. But what if this were not the case? What if your commute to work didn't require any mental faculties at all?

Imagine someone with an important job (a biotech researcher perhaps) who must spend an hour each day commuting to and from the lab. Suppose that, after the autonomous infrastructure is completed, his travel time is cut in half—thus earning him an extra half-hour of productivity each day. Now, consider bestowing this salvaged time across every day of the year and onto every biotech researcher in the country. And now consider similarly diminishing the commute time of every worker in America.

What could you accomplish in your own career if your commute time was cut in half?

Whatever the gains might be, they are surely substantial when considered on a nationwide basis. A 2013 blue paper by Morgan Stanley estimated that switching to self-driving cars could be worth $645 billion in worker productivity gains—primarily by reducing traffic congestion and assuming that a commuter could work for at least 30% of his road trip.

But even if self-driving vehicles do not immediately succeed in reducing the actual *duration* of the commute, they will surely be eliminating the need for mental output from the driver. Many workers would gladly accept a modest increase in their commute time if it meant that they didn't have to touch the wheel during the trip. If workers had the freedom to pop open a laptop or nap during their ride home, many would switch to a robo-taxi platform for this reason alone.

But we still have nine more reasons to go…

Reason 2: Self-driving cars don't crash

Cars kill people—a lot of people.

For many, the best way to convince them that the transportation sector is in dire need of an upgrade is to merely mention how very bloody our roads are. To put things in perspective, we can compare car deaths to war deaths.

United States military deaths (from the Revolutionary War of 1775 to our misadventures in the Middle East) have claimed around 1.1 million American lives. But over 3.7 million Americans died in motor vehicle fatalities from 1900 to 2019—meaning that our cars are *three times* deadlier than our wars.

Our worst year for US auto fatalities was in 1972 when 54,589 people died on the road. This number is almost identical to the *total* number of Americans who died in Vietnam—during our entire 18-year involvement.

So why are these numbers so high?

The National Highway Traffic Safety Administration (NHTSA) breaks down traffic fatalities into four groups:

- 32% of lives are lost in drunk driving related accidents
- 31% to speeding
- 16% to distractions
- 11% are due to driving in bad weather

In glancing at this list, we can see that almost all traffic fatalities are born of *human error*—i.e. the driver's inability to stop drinking, stop speeding, stop texting, or stop going out in unfavorable weather conditions.

Thanks to recent improvements in road and vehicle design, the current tally of US auto fatalities has fallen to around 36,000 per year. That's better. But pushing this number down to *zero* will probably require removing humans from the driving equation altogether. As driverless technology continues to improve, people will eventually come to understand that riding in a vehicle laced with a dozen digital eyeballs is much safer than riding in a vehicle equipped with only two.

Reason 3: Self-driving cars are environmentally friendly

The US consumes 7.5 billion barrels of oil each year. Around *half* of that oil is used for transportation fuel. Estimates by the EPA and the EDF claim that motor vehicles are responsible for 75% of carbon monoxide pollutants, and

35% of smog-causing air pollution. Transportation in general causes 27% of greenhouse gas emissions.

We pay an environmental cost for burning all this fuel. Cars emit a toxic cocktail of hydrocarbons, carbon monoxide, nitrogen oxides, and particulates that assemble to form the brown layer of smog that we have come to expect from our cityscapes. For those who live near major highways, traffic pollution has been blamed for a myriad of health issues including asthma, lung disease, heart problems, cancer, and even impaired cognition.

Thankfully, more stringent fuel standards and advances in engine design have substantially reduced tailpipe emissions over the last fifty years. In 2013, the EPA joyfully reported that:

Compared to 1970 vehicles, new cars, SUVs, and pickup trucks are roughly 99 percent cleaner for common pollutants.

When it comes to electric cars, things look even better. In reviewing the 2018 release of the EPA's power plant emissions database (eGRID2018), the Union of Concerned Scientists concluded that:

...the average electric vehicle produces global warming pollution equal to a gasoline vehicle that gets 88 miles-per-gallon.

That's a nice improvement over conventional autos. But electric vehicles aren't nearly as green as we'd like to think they are. For some perspective, your old Ford Fiesta was already capable of 37 miles-per-gallon. Electric cars moved that dial from 37 to 88. That's better, but not by an astonishing amount. We still have a long way to go before we can consider such vehicles to be truly "green."

As of 2018, the US electric vehicle market share is a minuscule 2.1 percent. But even if everyone started driving Teslas today, this still wouldn't solve all of our environmental problems. So long as our cars are still running on electricity that is generated via the burning of fossil fuels, then the convenience of travel will still come at the expense of the environment.

Cleaning up our power generation methods

For decades, critics have noted that the batteries of electric vehicles consume electricity that is generated via conventional means. In the US, the three most common sources of electricity are coal (30%), natural gas (34%), and nuclear (20%).

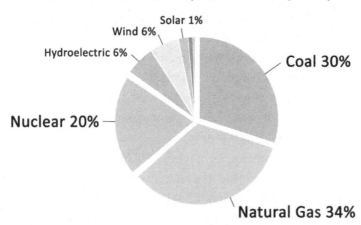

USA Net Electricity Generation (2016)

Solar 1%
Wind 6%
Hydroelectric 6%
Coal 30%
Nuclear 20%
Natural Gas 34%

Figure 12 - This 2016 chart by the US Energy Information Administration (EIA) reveals that 84% of US power generation comes from coal, nuclear, and natural gas power plants. "Green" power generation still lags far behind conventional methods.

Let's take a moment to describe each of these power generation methods now:

- **Natural Gas** is often hailed as a viable "clean energy" option. It does burn cleaner than other fossil fuel alternatives—producing less ash and fewer greenhouse gases. But it is a non-renewable resource. New natural gas sources must be continuously discovered and mined. Methods for retrieving natural gas from deeper deposits (e.g. "fracking") are controversial among environmentalists—who claim that the process pollutes the water supply and releases methane gas into the atmosphere.

- **Coal** burning power plants are dirty. Aside from the commonly cited respiratory illnesses—like asthma and lung cancer—coal toxins may cause heart disease, stroke, and even mental disorders. Modern-day coal power plants are much cleaner than their predecessors. But any power-generation method that spews bituminous smoke into the atmosphere should not reflect the standard for environmental-friendliness in the 21st century.

- **Nuclear** is by far the most controversial of all the power generation methods—due in large part to the widely publicized disasters at Fukushima, Chernobyl, and Three Mile Island. It also doesn't help that Hollywood loves portraying dramatic nuclear disasters in science fiction films. Still, the nuclear industry persists. Over the last decade, several startups have breathed new life (and new physics) into the mechanics of power generation via nuclear fission. Even the field of nuclear fusion—the long-awaited darling of futurist and clean energy advocates alike—continues to be on the verge of evermore exciting breakthroughs.

- **Solar** power is unique in its ability to simultaneously attract the interest of investors and consumers, while consistently failing to contribute anything substantial to the national power grid. In 2017, solar accounted for only 1.7% of global electricity production. But, though it's current contribution is tiny, solar has been steadily growing. In 1970, solar power costs $100 per watt. In 2009, the Arizona-based solar panel manufacturer *First Solar Incorporated* became the first company to lower this cost to 1 dollar per watt. More recent claims have touted even lower numbers of around 30 and 17 cents per watt. In the years to come, improvements in local energy storage and residential solar systems could be poised to disrupt the market. The number of solar installations has doubled every two years since 2000. In 2016, Elon Musk (co-founder of Tesla and SpaceX) acquired the solar panel manufacturer SolarCity for $2.6 billion—with the intent of integrating Tesla's highly-efficient lithium batteries into a complete residential and commercial solar solution. In sunny Australia (where rooftop solar is already installed on one in five homes) Tesla is building the "world's biggest battery"—scheduled to supply 129 megawatt-hours to the Australian grid.

In considering the preceding list, there is reason to be optimistic about the green future of solar and nuclear technologies. If power generation and battery manufacturing methods continue to improve then we will someday have plenty of clean energy to fuel the electric autonomous vehicles of the future. Recent solar advances have even compelled the Stanford economist and electric vehicle crusader Tony Seba to audaciously proclaim:

By 2025 everything that moves...will be electric.

Fleet Asset Utilization

Even if power generation methods fail to significantly improve in the years to come, autonomous ridesharing vehicles are still our greenest and most efficient option—simply because their expected asset utilization is so much higher than that of conventional autos. Recall again that your car will spend 96% of its life sitting in a parking space. That's bad. But when you consider that most cars have five seats—and that *four* of these seats are typically empty during road trips—then the waste and inefficiencies of personal car ownership are made all the more evident.

Robo-taxis don't sit in a parking space after dropping off each passenger. Instead, they return to the road to pick up a new rider. If you take a moment to ponder the implications of this schema, then you may arrive at an encouraging realization. Increasing the asset utilization of passenger vehicles would mean that the nation wouldn't need to build so many cars. This insight prompted Seba to predict that the nation's car fleet will shrink by an incredible 80% over the next ten years.

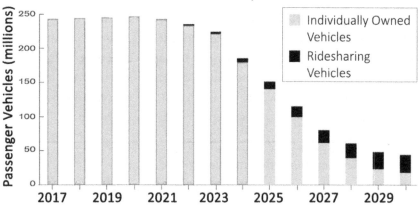

Figure 13 - Tony Seba predicts that, over the next 10 years, the number of cars in operation will shrink by 80%. Only half of all passengers will own their own car. The rest will use ridesharing robo-taxis.

We have to pull a lot of resources out of the earth to build something as complicated as an automobile. The fact that these assets spend so much of their lives doing *nothing* is an environmental tragedy. In the coming autonomous world, we won't need to devote so many resources to the construction of such large and sedated vehicle fleets. A smaller national car fleet will reduce the environmental impact of transportation as well as help to decrease urban traffic congestion. For environmentalists (or even for people who just hate wastefulness) this is promising news.

Reason 4: Self-driving cars will end traffic congestion

Americans spend about 7 billion hours each year sitting in traffic. In attempting to estimate the value of these lost man-hours, INRIX Research arrived at a dollar amount of $305 billion annually. This figure includes direct costs to the driver (in fuel and time) as well as indirect costs to business productivity.

In addition to financial losses, traffic jams incite personal frustration. The tedium of sitting in rush-hour traffic can lead to anxiety, depression, road rage, or aggressive driving—which results in yet more traffic accidents and more congestion.

What causes traffic jams?

Half of all traffic jams occur as a result of road work, bad weather, or traffic "incidents"—in the form of auto collisions, complicated lane merges, irregular braking, and rubbernecking. But the other half is the result of the complexities that arise when we ask an error-prone human to drive a vehicle on a road filled with other error-prone humans. Mistakes will be made—even in ideal conditions.

Recent traffic flow experiments have managed to simulate car pileups on roads that are completely free of obstructions. So-called "phantom traffic jams" can result from the tiniest driver blunder—like when car brakes are tapped just a bit too hard, or when a turn signal is engaged for a bit too long. In heavy traffic, such minor events often swell into a tumultuous complication. The mistake of one driver will ripple down the line of cars—eventually evolving into a full-fledged and self-sustaining traffic jam—which could last for several hours.

It is tempting to conclude that traffic congestion could be eliminated if we were to just build more roads. But even if the money and space were available for unbounded construction, merely adding more lanes would not eliminate traffic jams—because traffic jams are not solely the result of too many cars on the road. Instead, traffic jams are the result of too many humans at the wheel.

- People have different driving styles and different dispositions by which their brain tailors a solution to solve each navigation puzzle as it permeates their narrow field of view. The strategies employed by one driver may be incompatible with those of another.
- Drivers have a limited amount of cognitive resources from which to draw upon. They are limited in the number of decisions that they

can make per second, as well as the number of objects that they can keep track of.

- Reaction times, visual acuity, and depth perception will vary between drivers. They also vary throughout the day—in the evening, driver fatigue inevitably sets in and hinders one's ability to perform these functions. Even the most adept of us are not capable of error-free driving for more than a couple hours each day.

Freeing automobiles from their traffic quagmires will require an operator with exceptional navigation abilities, infinite stamina, and enough grace to courteously share the road with every proximate car. Humans will never be able to satisfy these requirements. But fortunately, we have an alternate candidate for the job:

- Autonomous vehicles never get frustrated, bored, or tired.
- They don't break the speed limit nor do they slow down to rubberneck an accident.
- Autonomous vehicles don't ride the brake, they don't tailgate, and they don't get road rage.
- They also don't drink or get distracted. Nor do they ever get lost.

We can only eliminate traffic jams by eliminating the convergence of human propensities that incite their formation. This is why mankind will ultimately surrender the road to the more capable hands of a future self-driving A.I.— at least someday.

Swarm Intelligence

Aside from their ability to play well with others, self-driving cars will be in constant contact with traffic computers—receiving second-by-second updates about changing road conditions. This will allow them to alter their route based on things like weather, police activity, traffic accidents, or changes in congestion levels. Future autonomous vehicles may even utilize swarm intelligence. Proximate vehicles will create an impromptu data network that tracks each car's vector, size, and intended route—thus enabling neighboring vehicles to anticipate and compensate for any navigational divergences. Human drivers currently accomplish such communiqués using turn signals, headlights, and eye contact. The sluggish pace of our morning commute is the direct result of such limited signaling apparatus. So long as our vehicles are piloted by people, it is unlikely that travel speeds will ever increase beyond the present average. However, if every vehicle on the road is piloted by A.I., then more impressive speeds could safely be achieved.

Reason 5: Self-driving cars are fast

Formula One racecar drivers attain speeds of around 115 miles-per-hour during competitive racing. On some segments of straight and uneventful highway, autonomous vehicles might one day be capable of a similarly impressive pace. Stanford mechanical engineering graduate Nathan Spielberg is one of several academics currently working to develop autonomous cars that are capable of supercar speeds. Using neural networks to achieve high-performance driving, his autonomous cars have been tested at Thunderhill Raceway in Willows, California, and were able to achieve lap times that were competitive with trained drivers. Commenting on the future potential of high-speed AV technology, Spielberg said:

We want our algorithms to be as good as the best-skilled drivers—hopefully better.

If future robo-taxis can someday reach cruising speeds that are on par with professional racecar drivers, then average commute times will decrease and much longer commute distances will be possible.

Reason 6: Self-driving cars are fuel-efficient

Internal combustion engines typically have over 2,000 moving parts and can only achieve an efficiency of around 30% on the highway. Most of the gasoline they consume is wasted—converted to heat and smoke. Electric motors, on the other hand, may have as few as 20 moving parts and can ideally convert over 90% of their incoming energy into mechanical power.

When it comes to running these machines, ICE engines are burdened with their dependence on fossil fuels. Aside from the cost and environmental impact of the petroleum extraction process, additional work is required to create the final gasoline product—which must be refined, transported, and distributed to each gas station in America—all 121,998 of them.

Electric vehicles offer a superior alternative to gas stations. Their charging stations do not require constant refueling nor must they reside atop a 12,000-gallon tank. Instead, charging stations need only be plugged into the city power grid. Meaning that they can be constructed almost anywhere (including parking structures and home garages), and they do not require tanker trucks to make periodic deliveries. As Tony Seba likes to point out, "electrons are cheaper to transmit than atoms."

Along with fuel transportation savings, there is yet another (less obvious) benefit to be garnered in the use of autonomous vehicles. Eventually, when the vector of every car on the road is governed by an all-knowing traffic management computer, then algorithms can be designed to pursue the most efficient application of acceleration and deceleration throughout the vehicle's trip. That is to say, autonomous vehicles don't have a "lead foot."

When humans drive cars, they tend to stomp on the accelerator and the brake too hard and too often, because they are unable to anticipate how much fuel (or how much braking) will be required to adequately move the car to the next stoplight. But future autonomous vehicles will be perpetually aware of all proximate cars and all upcoming driving scenarios. This means that they will be able to predict exactly how much energy will be needed to move the car along its journey at optimal efficiency.

For an extreme example of just how efficient an optimized acceleration schema can be, consider the "Hypermiling" phenomenon. In 2015, San Diego residents Wayne Gerdes and Bob Winger drove their Volkswagen Golf across each of the 48 US contiguous states in just 16-days. During the journey, they averaged 81 miles-per-gallon and spent less than $300 on gas—thus earning them a Guinness World Record for fuel economy in a non-hybrid car. Hypermilers accomplish such feats (in part) by scrutinizing over every tap of the accelerator and brake. As their car rolls down hills or approaches stop signs, drivers attempt to salvage every bit of potential energy in their vehicle—gracefully accelerating or anticipating upcoming decelerations as needed.

Of course, devoting this level of attention to each morning commute would not be practical for casual drivers. However, such minute tunings will be easy for the autonomous vehicles of the future; their onboard computers will foresee every upcoming turn, hill, and stoplight. So the car will be able to adjust its required energy output accordingly—guaranteeing the speediest and most fuel-efficient route to each destination.

Reason 7: Self-driving cars are less expensive to maintain

In 2019, the consumer credit company Experian reported that the average new car price is around $34,000. Smaller cars (like the Toyota Camry and Honda Accord) can be purchased in the low 20,000s. But the average is driven up by America's thirst for gigantic vehicles—like the Ford F-150 (MSRP: $28,025 - $64,515) and the GMC Sierra (MSRP: $29,400 - $56,800).

Shiny new cars become "money pits" shortly after they roll off the lot. And to keep these vehicles rolling, car owners must pay for:

- Fuel
- Insurance
- Financing
- Tires
- Scheduled Maintenance
- Licensing
- Registration Fees and Taxes

Recall that AAA estimated that the cost to operate a medium-sized sedan is $9,282 annually. But maintenance fees increase as cars age. Each successive trip to the mechanic requires the purchase of increasingly mysterious parts and services. After the first 25,000 miles on the road, expenses begin to ramp up. After 75,000 miles, costs double and they continue to gradually rise throughout the life of the car.

Mileage	Total Maintenance Costs per 25k Miles
0 - 25,000	$1,400
25,000 - 50,000	$2,200
50,000 - 75,000	$2,000
75,000 - 100,000	$2,900
100,000 - 125,000	$4,100
125,000 - 150,000	$4,400
150,000 - 175,000	$4,800
175,000 - 200,000	$5,000

Figure 14 - The online auto maintenance company YourMechanic.com examined their work history logs to compare mileage versus maintenance costs.

In the coming world of ridesharing robo-taxis, nobody will own a car. So none of us will be directly paying for any of the above-cited items. Instead, the cost to purchase and operate a robo-taxi will be bared by the taxi fleet owner. And, in turn, distributed upon the thousands of riders who will share the taxi's seats.

Reason 8: Self-driving cars don't need to be parked

Because personal automobiles are used so inefficiently (and because they remain stationary for the vast majority of their existence), we are forced to devote much of our lives to the coddling of these lackadaisical two-ton behemoths.

In a typical US city, parking spaces may account for over one-quarter of developed urban land—providing an estimated average of 3.5 parking spaces per car. In congested cities, especially those built before the invasion of the automobile, hunting for a parking space is a time-consuming chore. A 2017 study by the Washington based traffic analytics firm INRIX, estimated that Americans spend 17 hours each year looking for a place to park. In wasted time, fuel, and emissions this computes to a cost of $345 per driver (or $72.7 billion nationwide). Another $20 billion each year is spent overpaying for metered parking spots.

Parking in the Garage

Equipping each home with a personal garage helps to alleviate the pain of *parking space hunting*. But the remedy comes at a cost. In densely populated urban centers, there isn't enough room to build a garage for every apartment unit. In the suburbs, two-car garages are a necessary prerequisite for all residences. But they occupy a sizable amount of precious lot space—taking up around 576 square feet, along with a couple hundred additional square feet for an adjacent driveway. Moreover, the aesthetic charm of a home is often ruined by the addition of a garage—which often sports an overbearing roofline and a barren vertical fascia to accommodate the massive windowless planks that form its door. When one drives down the avenues of suburbia, these unfortunate "car guestrooms" are typically the most prominent architectural ornaments on display.

Parking on the Curb

Car owners who don't have garage access are forced to ditch their ride along the curb that rests in front of their homes or apartment buildings. Weeds of parked cars have lined our city streets for so long, that we have begun to think of them as permanent sidewalk fixtures. Thus far, the most commonly employed solution to the parking problem has been to devote ever more acres to parking lots, or to construct increasingly taller and more sophisticated parking structures. But these concrete monoliths require millions of dollars to build and maintain. And even when they are successfully erected, they stand only as monuments to engineering inefficiency—

needlessly extravagant warehouses devoted to the storage of infrequently used machines.

Fortunately, a better solution is coming. And we'll soon be able to do away with residential parking structures and home garages altogether.

Since robo-taxis don't need to be parked after completing a trip, they do not require on-site parking facilities at every destination they visit. Once they drop off a passenger, they simply return to the road to find another one. When their shift is over, the taxis will drive themselves back to home base— a strategically positioned facility built for maintaining autonomous vehicle fleets. Here, the cars will get a charge, a car wash, and mechanics will take care of any needed maintenance—readying the vehicles for tomorrow's morning rush.

When such an autonomous infrastructure is finally in place, the way we design our city blocks and neighborhoods will have to be reevaluated.

- Houses and apartment complexes will no longer need to devote so much land to the garaging of automobiles. Instead, homeowners will be able to convert their existing garages into game rooms, guest rooms, workshops, or even rentable space.
- Commercial buildings can be constructed without massive multi-level parking structures—the tenants and clientele won't need them anymore.
- And retail outlets, hotels, and shopping malls need no longer surround themselves with acres of black asphalt. Instead, a queue of robo-taxis will drop-off each customer at the facility's front entrance.

Instead of wasting precious space on parking lots, all urban buildings can be constructed with drop-off zones—street-accessible areas that will provide robo-taxis with space to rapidly offload their human cargo. We'll be talking about such dropzones more in Chapter 7.

Reason 9: Self-driving cars don't need to stop at intersections

Riding in a fast car is nice. But high-speed electric motors will not be the primary time-saving apparatus of the coming autonomous age. A simple *lack of speed* is not the reason that your trip to the shopping mall took so long last Sunday. Rather, most driving minutes are lost because of the dozens of stop signs and traffic signals that impede the path between your origin and your destination. Each time a car encounters an intersection, it must slow to a stop and wait for its turn to proceed. Such bouts of braking and acceleration waste energy, diminish passenger comfort, and increase travel time.

Figure 15 - The Yan'an East Road Interchange in Shanghai, China. (Denys Nevozhai)

To bypass these impediments, cities construct *stack interchanges*—highway junctions comprised of swooping scions of multi-graded roads. These massive concrete Gordian Knots do eliminate the need for cars to halt their forward progress when turning. But the benefit comes at great cost—such structures require millions of dollars and several years' worth of planning and construction. And, due to their immense size, they can only be built atop high-trafficked junctions. For the majority of intersections (of the type that one might encounter on the way to work each morning), this method can't be used.

The only other solution to the *intersection problem* has been to demand that each car heed to the authority of a hanging traffic signal. This strategy causes us to waste precious minutes of our lives—tapping a motionless steering wheel, practicing stoicism, and waiting for a series of red lights to turn green. It's an all-too-familiar quagmire that has slowed the wheels of progress since the inception of the automobile. But thankfully, the adoption of an autonomous infrastructure might allow for technology that will help to get our tires moving again.

Researchers at MIT, the Swiss Institute of Technology, and the Italian National Research Council have suggested that we might replace traditional traffic lights with a new "slot-based" intersection. In their schema, an approaching autonomous vehicle would broadcast its desired direction to a traffic computer. The system would in turn dispatch a "slot" (a time) at which the requesting car may enter the intersection. Based on this time frame, the incoming car would alter its speed—such that it will execute its turn at precisely the right moment—when the heart of the intersection is free of traffic. Theoretically, the car should be able to avoid colliding with other proximate vehicles—thanks to the watchful guidance of A.I.

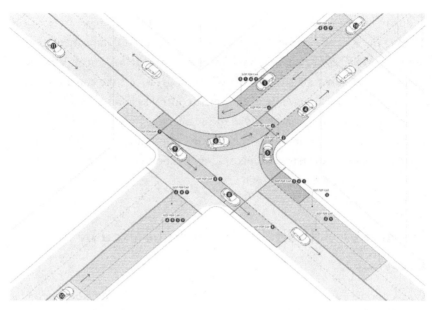

Figure 16 - In this slot-based intersection from the MIT Senseable City Lab, incoming vehicles ping the traffic management system and request a "slot" (a time) at which their desired route will be free of other cars.

If researchers succeed in perfecting such a system, it could potentially do away with traffic signals as we know them—thus relinquishing the need for automobiles to make so many stops while en route. Much of the energy used in road travel is expended during the acceleration and braking of the vehicle. But if braking is rarely required, then cars need not waste this energy.

Such an intelligent choreography would markedly shorten the duration of every car trip. To appreciate how valuable this invention could be, consider what your current commute would look like if you never had to make a single stop—i.e. if every traffic light at every intersection turned "green" just as soon as your car approached it.

- Imagine how much time could be saved per day, per year, or per lifetime?
- What if every single working American also saved this much time each morning?

If such a device replaced every traffic signal in the country, the potential gains to national productivity would be astonishing.

Industry investors are sometimes mesmerized by the sexy red curves of the coming fleets of autonomous vehicles. But the successful implementation of *autonomous traffic signals* will be a similarly momentous achievement. While self-driving A.I. is sure to hasten the wheels of progress, it is the concinnation of autonomous vehicles *and* intelligent traffic control systems that will slam the accelerator pedal to the floor. Of all the ancillary technologies that will arrive on the coattails of autonomous vehicles, the importance of intelligent traffic signals cannot be overstated. This sole innovation could be responsible for saving billions of man-hours each year.

Reason 10: Self-driving cars can be used by everyone.

Conventional cars can't be operated by every citizen. Aside from the great expense of personal car ownership, obtaining a driver's license requires a test of intellect and physical dexterity. Not all of us are capable of passing this test.

- Cars can't be driven by the very old or the very young.
- Cars can't be driven by anyone consuming alcohol, marijuana, or disorienting prescription medications.
- Cars can't be driven by the blind, the sick, or the physically handicapped.

Even for those of us who do manage to score a driver's license, none of us can be expected to drive responsibly every single time we get behind the wheel. If drugs and alcohol don't cause our minds to wander, sheer boredom often does. Most people don't particularly enjoy driving. During long trips, we attempt various techniques to combat the hours of tedium. We listen to music, or to the radio, or to audiobooks; we try to avoid the deep-rooted neurophysiological temptation to examine our cellphones—though, few of us are ever able to resist for long.

No human driver can consistently maintain a satisfactory level of focus during every single second of a road trip. It would be irrational to assume that this would even be possible. And our insistence on maintaining this pretense is part of the reason why so many lives are lost to driver fatigue and distraction.

Self-driving cars are not burdened with such human foibles. Navigation computers never get bored. All they do is drive. Flighty robo-taxi passengers will be free to let their minds wander for the duration of the trip. And for those who cannot drive, autonomous technology may provide an even greater gift.

In 2015, Nathaniel Fairfield, the Principal Engineer of Waymo (Google's self-driving technology company), asked Steve Mahan if he'd volunteer to take a solo trip in their prototype self-driving vehicle. The trip would be significant because he would be the first passenger to ride in a self-driving car on a public road—without a police escort or a backup human driver. Steve Mahan is legally blind and hence, is representative of the very type of person who could benefit the most from this technology.

After completing his October ride through the streets of Austin, Mahan commented:

This is a hope of independence. These cars will change the life prospects of people such as myself. I want very much to become a member of the driving public again.

Autonomous technology will free us from the danger and tedium of driving. This is good. But for those who cannot drive, this technology is all the more crucial; self-driving cars are the instrument that will help to restore their personal independence and freedom.

Ch. 3: Mobility as a Service (MaaS)

In Chapter 1, we described an autonomous robo-taxi with a four-seat configuration that is ideal for carrying groups of solo riders—the sort typically found during morning commutes. Our taxi used an in-line seating configuration which discouraged socializing. While most blurry-eyed Monday morning commuters might prefer such an arrangement, more diverse seating options will be available in other vehicle platforms. In the following diagram, three additional configurations have been placed next to our initial robo-taxi design:

Seating Configurations for
Autonomous Ridesharing Robo-Taxis & Shuttles

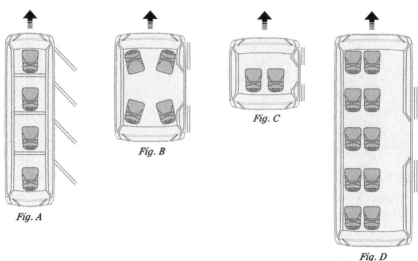

Figure 17 - To accommodate varying social and logistical situations, several types of seating configurations will be utilized in future autonomous platforms.

Note that each vehicle can accommodate a different number of passengers, and each one facilitates a different degree of social interaction. Unlike the antisocial arrangement of the seats in our first robo-taxi design (*Figure A*), the layout in our family-friendly model (*Figure B*) features four swivel seats which encourage conversation among riders. Such taxis might be used by friends, coworkers, families, or people with more gregarious personalities. This configuration is very similar to that of the Mercedes Benz F-015 prototype shown in the following image.

Moving on to *Figure C*, we see a simple two-seat robo-taxi. Readers will note that this model is very similar to the Google "Firefly" prototype discussed in Chapter 1. Such minicompacts are possibly representative of the type of *city car* that will be utilized for brief trips around town—which comprise the majority of daytime outings.

Finally, in *Figure D*, we have a much larger ten-seat shuttle bus. These vehicles may be best suited for Demand Responsive Transport (DRT)—sometimes called Dial-a-Ride Transit (DART). Like a DRT vehicle, our autonomous shuttle might alter its schedule based on demand, rather than stick to a fixed route, as the city bus does. And like Uber Express Pool, riders may be asked to meet the shuttle at a street corner that is a "short walk" from their origin. If multiple pick-ups are requested in the same general vicinity, the navigation computer might create an impromptu "virtual bus stop" and direct each rider to meet the shuttle there—so that it can pick up both passengers at the same time.

Other Amenities

Aside from the four platform configurations described above, the autonomous vehicles of the future will undoubtedly offer countless other seating arrangements and amenities.

- Some vehicles may be designed with very young children in mind. Perhaps featuring smaller seats or even protected passenger compartments—equipped with doors that cannot be opened until the car has made it to grandma's house, and the child occupant is fetched by a designated adult.
- Some shuttles may offer facilities for passengers with special needs—like handicapped access ramps, wheelchair storage, or door-to-door passenger assistance.
- Some taxis may have age restrictions. For example, taxis for senior citizens may refuse to allow riders under 65 years old onboard. Cars catering to a more *adult experience* might not open their doors to anyone under 21. And might even sell alcohol on board.

Rider Rating Systems

Most of us are familiar with Uber's rating system. It's that little screen that pops up on your phone following the completion of each Uber ride—prompting you to rate your driver's level of skill and courtesy. Given the utility of these existing systems, future autonomous ridesharing companies will probably provide their passengers with a similar sort of evaluation app.

- For example, just as you can use the current Uber app to "blacklist" drivers who rub you the wrong way, it may be possible to blacklist your fellow ridesharing commuters as well. You might be notified if the undesirable party is scheduled to be in the same robo-taxi as you. If they are, you'd be given the option to select a different ride.

- On the other hand, riders might also be able to maintain a "whitelist"—a list of fellow friends, commuters, or coworkers who they do enjoy traveling with, and who they wouldn't mind sitting next to on the way to work.

- Perhaps ridesharing apps could even be paired with dating apps. Thus allowing romantic potentials to share a car during their commute—giving them a few moments to chat each other up on the road. Later, when the trip is over, the app would ask each participant if they were interested in another encounter.

Assumedly, many taxi features and seating arrangements will be tried by the enterprising startup companies of the future. Most proposals will fail. But, we will eventually settle upon a set of foundational services that will be widely available in autonomous ridesharing vehicles throughout the country. When such an infrastructure is in place, personal car ownership will be uneconomical. Instead, autonomous taxis and shuttles will carry passengers for the majority of short-range trips.

But what about long-range trips?

Scheduling trips via MaaS apps

Longer trips are complicated because they often require multiple types of vehicles. Typically, passengers must first utilize smaller types of transport (like taxis and subways) and then progress up to larger forms of mass-transit (like planes, trains, and ships). Often, each vehicle must be independently coordinated, booked, and scheduled. During the journey, passengers are required to stop at several check-in counters, interact with independent payment points, and make complicated transfers between vehicles and terminals.

To illustrate this pandemonium, let's describe a typical journey. Consider a student who needs to travel from his home in Queens, New York to attend a conference in Los Angeles. Similar treks have commenced thousands of times each day for decades—so you might assume that the process is refined by now. Yet, even simple plane trips are laden with points of friction. Consider his itinerary:

- First, our student must independently arrange at least three modes of transportation—one for the airline, one for the trip to the airport, and one for the trip from the airport to his hotel.
- On the morning of his flight, he must haul his suitcase from his bedroom to the curbside, and then hoist it into the taxi's trunk.
- After arriving at the airport, the taxi driver must locate the appropriate outbound terminal and then temporarily park at the curb to drop off the student and his luggage.
- Then, our student must haul his suitcase to the airline check-in counter—where the attendant tags his bag and makes it disappear on a conveyor belt. Assuming the student has prepaid for his ticket, the attendant will then issue a boarding pass.
- Our student then proceeds through the security checkpoint, walks through the terminal, and arrives at the concourse to await boarding.
- Hours later, when his plane arrives in Los Angeles, our traveler must endure the comical inefficiency of the baggage carousel—waiting for his suitcase to come tumbling down its metal ramp.
- Once he manages to snatch up his bag, he must hire a third vehicle. Perhaps he'll use a shuttle to get from the airport to his hotel.
- Finally, after the shuttle arrives at his destination address, it will require one last valiant effort to haul his suitcase from the shuttle to the hotel's check-in counter. And then up the elevator to his room.

In perusing the above list, it's easy to see why air travel is such an anxiety-inducing enterprise. Though our above-described trip is as simple as they come, it still required three independent financial transactions and the coordination of three vehicles (the taxi, the airplane, and the shuttle). Aside from all of the typical air travel annoyances (e.g. the bad food, the lack of legroom, and the in-flight turbulence), it's the mad rush through each waypoint that is the most anxiety-inducing part of the journey. Thankfully, a new system is emerging that should make the trek easier.

What is MaaS?

Most people are not yet familiar with the term "MaaS" (Mobility as a Service). It refers to route-planning software that consolidates multiple forms of transport into a single scheduling and payment gateway. In a MaaS schema, each leg of the journey is managed (and paid for) via a single application. To move the passenger from Point A to Point B, the computer schedules each vehicle that will be needed to complete the trip. From the passenger's perspective, all he needs to do is enter his destination address and follow the instructional prompts on his cellphone as the app directs him to each waypoint.

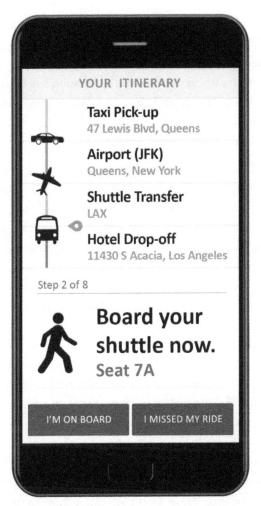

Figure 18 - In this MaaS app, three different types of vehicles are managed via a single application gateway.

Because the software is aware of the travel characteristics and time constraints of each vehicle in the passenger's itinerary, each transfer can be intelligently coordinated. Vehicles can be scheduled to pick-up and drop-off a passenger at the most efficient moment—granting the traveler with the ideal amount of time to exit one vehicle (like his taxi) and board another (like his departing train or airplane).

In a MaaS schema, minimal check-in procedures are required. Your cellphone acts as your boarding pass and because you have already used the app to prepay for every vehicle needed for the trip, further transactions are not required while en route. So passengers won't be purchasing any additional tickets nor negotiating with any unscrupulous taxi drivers. Coordinating existing modes of transport into a single application will ease the burden of travel. But this load will be even lighter once MaaS apps are integrated with autonomous vehicles.

MaaS + Autonomous Vehicles

In the future, all vehicles will be autonomous vehicles. And thus, all vehicles will be connected to the internet. This connectivity will make it possible for MaaS applications to coordinate every type of vehicle needed to complete a passenger's journey. Smaller vehicles (like robo-taxis) will be able to intelligently intercept mass-transit vehicles (like planes, trains, and ferries) and facilitate the loading and offloading of passengers and baggage. Below, we'll describe *three* travel features that such an infrastructure will make possible.

Feature 1: Vehicle-specific coordination and wayfinding

Because a MaaS app will be able to coordinate with every vehicle that is to be used during the journey, it will help to curtail the logistical tedium of travel for the passenger. The app will be receiving constant updates about the state of each vehicle, and will be feeding the traveler with step-by-step directions at every waypoint. From the rider's perspective, all he has to do is follow the steps on his phone. This feature will be particularly beneficial in large airports and terminals—where the app will not only be aware of the passenger's departure time, but also of how he is to navigate through the building itself.

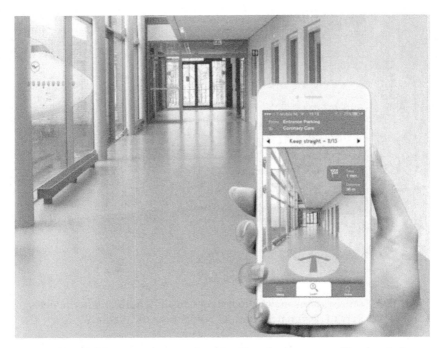

Figure 19 - This concept app by the Dutch company *Eyedog* helps a passenger navigate through an airport terminal by rendering an arrow icon on the incoming camera feed from her cellphone.

Feature 2: Transit stations will be retrofitted to accommodate incoming robo-taxis

As robo-taxis become the primary mode of transportation, transit stations will someday revamp their platforms to accommodate rapid passenger drop-offs. Ideally, such stations will place their platforms near the road so that incoming autonomous vehicles can release their passengers as close to the outbound mass-transit vehicle as possible.

Figure 20 - Some stations may be redesigned to allow autonomous vehicles to carry passengers to designated drop-off points near the platform.

Feature 3: Autonomous luggage routing

While still relatively unknown, the luggage-forwarding industry has been steadily growing for the last two decades. The largest US company (Boston-based *Luggage Forward*) has been shipping suitcases for air travelers since 2004.

Their process is simple:

- First, a passenger schedules a luggage pick-up online—usually for the same date as his outbound flight.
- Then, Luggage Forward retrieves the customer's suitcase from his home address and prepares it for shipping.
- Conventional delivery companies (like FedEx, UPS, and DHL) are used to send the luggage ahead of the passenger.
- When the passenger finally does arrive at his hotel, his luggage is there waiting for him.

Luggage Forward's business model offers a glimpse into the future of air travel. But, because such companies rely on conventional shipping methods, their model suffers from all of the typical logistical problems that every other mail courier must contend with. In the next chapter, we will describe how autonomous courier vehicles will help to alleviate package pick-up and drop-off difficulties. When such obstacles are finally surmounted (when the service of delivery is automated), then people will no longer be hauling suitcases to the airport nor dragging them to the airline's check-in counter. Instead, the process of shipping a bag from Point A to Point B (luggage or otherwise), will be entirely handled by machines. Autonomous courier vehicles will retrieve the suitcase from your home and inject it into a grand clockwork of package-delivery machinery—which will move the bag along to its final destination. Such conveniences will eliminate many of the pain points that are so commonly associated with air travel. There will be:

- No more suitcase check-ins at the airline counter.
- No more waiting at the baggage carousel.
- No more boarding passes.
- No more airport parking lots.
- No more getting lost in stations or terminals.
- And, no fast-paced pick-up and drop-off dramatics at the airport curbside.

For medium and long-haul flights, the airline industry would benefit from autonomous vehicle integration for both baggage-handling and passenger transport to and from the airport. However, when it comes to short-haul flights the industry may be in for a disruption.

The Disruption of the Airline Industry

In 2009, the FAA estimated that US commercial aviation supports 10 million jobs and accounts for 5% of GDP—contributing $1.3 trillion in annual economic activity. Airlines move 58,000 tons of cargo and 2.4 million people each day. Perhaps because airplanes are capable of flying so fast, we tend to focus on their impressive cruising speed and forget about the many inconveniences that are encountered in the space between our doorstep and the airplane seat.

In air travel, precious time is lost:

- at the ticket counter
- at the security checkpoint
- during the wait for boarding
- during the layover between connecting flights
- during delays on the tarmac
- and during the wait at the baggage carousel

Additionally, air travel always consists of at least three vehicles:

1. A ride from your home to the airport
2. The plane ride itself
3. And another ride from the terminal to your destination

Given the size of most airports, moving walkways, people-movers, lite rail cars, and additional shuttles are often required to complete a trek through the terminal. But once you're finally on board, the annoyances don't stop. Airline travelers must contend with:

- screaming toddlers
- chatty neighbors
- seats that don't fully recline
- a lack of legroom
- bouts of turbulence
- and horrifying crash scenarios

When all of these hang-ups are taken into account, the actual travel time is *always* much longer than the time indicated on the passenger's ticket.

This is one of the reasons why short-haul domestic flights (which United Airlines defines as any flight under 700 miles) have been on the decline for over twenty years—down by almost 30% since 2000. A troubled airline industry, increased federal taxes on air travel, and rising prices all contribute to the fall. But the sheer tedium of travel-by-plane may be the biggest culprit. The public has long suffered the pains of air travel because the alternative modes of long-range transport (busses, boats, and trains) have always been worse. However, when it comes to short-haul domestic flights, that may be about to change.

Long-Range Self-Driving Cars

Self-driving tech companies are quick to proclaim that (someday), their technology will allow you to "sleep while you drive." The Volvo 360c prototype caters to this desire—famously depicting a beautifully rendered autonomous vehicle, speeding down the highway with a single reclining passenger on board.

Given the many airport inconveniences that we discussed in the previous section, auto manufacturers are aware of the potential for autonomous technology to disrupt the airline industry. As Volvo's senior vice president of corporate strategy Marten Levenstam said:

Domestic air travel sounds great when you buy your ticket, but it really isn't. The [autonomous Volvo] represents what could be a whole new take on the industry. The sleeping cabin allows you to enjoy premium comfort and peaceful travel through the night, and wake up refreshed at your destination. It could enable us to compete with the world's leading aircraft makers.

If auto manufacturers manage to convince airline passengers to switch from planes to cars, then they will be tapping into a new market of moneyed travelers. Even if they only manage to overtake a modest portion of short-haul flights, such a disruption would still be impressive—billions of travel dollars would be redirected from the air to the ground. But will future self-driving cars really fulfill Levenstam's pledge to provide passengers with "premium comfort and peaceful travel through the night?" What sort of vehicle would be capable of long-range transport of this sort?

The Nightliners

Many will undoubtedly find Volvo's glossy renderings enchanting. And it's tempting to think that self-driving cars will soon be sprinting us through the night under the watchful guidance of A.I. But will such a feature really replace air travel?

Jetting down a moonlit road in a self-driving Volvo 360c would probably be better than driving your old Honda Civic all night—gripping the steering wheel with jittery fists, bloodshot eyes, and a mug of coffee from 7-11 in the slide-out cup holder. But the limitations of Volvo's new design would be revealed as soon as you needed to pee.

More importantly, while a Boeing 747 is capable of hauling 366 passengers, the Volvo 360c only carries one. So you would need a lot of Volvos to compete with the carrying capacity of one jetliner. Thus, it is unlikely that this particular type of vehicle will ever be able to significantly disrupt the airline industry. But what if we added more beds?

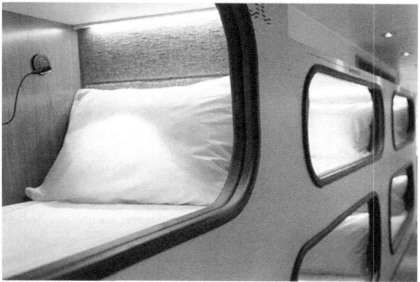

Figure 21 - For $120, the California sleeper bus company "Cabin" offers an overnight route from Los Angeles to San Francisco.

A "sleeper bus" (or "nightliner") is a long-range bus that is equipped with passenger beds and bathroom facilities. They have been in use throughout Europe and Asia for decades, but have not been commonly used in the US until recently.

Founded in 2016, the California-based sleeper bus company "Cabin" currently offers an overnight trip from Los Angeles to San Francisco. Cabin passengers meet at a designated station, stow their luggage on board, and then crawl into a pod-like enclosure for the night.

Cabin is expanding—adding two additional buses to their California fleet. In 2017, the company secured $3.3 million in seed financing and plans on someday going national. Reportedly, many passengers have come to prefer the "Cabin experience" to that of air travel. However, the technology is far from perfect and passenger complaints are easy to find online. Let's take a moment to consider five flaws in this "drive while you sleep" transportation model.

1. First, note that the Cabin sleeper bus doesn't pick you up at your front door. To get on board, you'll need to catch a ride to their bus stop in Los Angeles. And when the bus arrives in San Francisco, you'll need to fetch yet another car ride to complete your journey. Because the Cabin model still relies on multiple types of vehicles, the solution doesn't reduce the number of passenger transfers required to complete a trip.

2. Cabin's unique selling point is that their passengers can sleep during the eight-hour trip to San Francisco. When the bus arrives at 7:00 a.m., the passengers wake up "refreshed" and ready to start the day. (At least, that's what their ad copy says.) While it seems plausible in theory, experienced travelers know that this is an unlikely outcome. Truly restorative sleep requires more than a flat cushion. When you try to sleep in an unfamiliar environment, the brain's left hemisphere remains in an alert mode—its threat-detection apparatus is active throughout the night. The phenomenon is known as the *First Night Effect* and it's the reason why hotel rooms (despite their lavish amenities) often don't succeed in providing us with quality sleep. This condition is further irritated by the many potholes on the highway between Los Angeles and San Francisco. Cabin is installing their own patented "bump-canceling bunk beds" to try to compensate for the shoddy state of Interstate 5. And their GPS software is programmed to direct the driver to avoid rutted portions

of road. But, even with such countermeasures in place, a complete night of blissful rest will be difficult to achieve for most passengers.

3. Earlier versions of Cabin's passenger pods didn't have enough room for people to sit up in bed. Maintaining a horizontal position is easy if you're sound asleep for the entire trip. But if you'd like to sit up and stretch your back, you can't. The original Cabin sleeping pods weren't much bigger than coffins.

4. Even more discouraging is Cabin's lack of showers. After eight hours in a public bus bed, starting a day at your San Francisco office (sans shower) might not put you in your most pleasant and confident state. For many, the ritual of morning grooming is just as important as a good night's sleep.

5. Finally, all of these problems are compounded by the fact that a Cabin trip costs $120. The equivalent airline ticket only costs $97. Additionally, the total Cabin travel time is over eight hours in duration. Which is about six times longer than the 1 hour and 22 minute flight time that the airlines provide.

In consideration of the preceding, this "sleep while you drive" mode of travel (that is so often touted by self-driving car enthusiasts) is unlikely to adequately replace most air routes. While vehicles like the Cabin bus and the self-driving Volvo concept car are interesting, they are flawed specimens—representative of the transitional state of autonomous transport technology.

However, there may be room for a different sort of vehicle—one that combines the convenience of a door-to-door taxi service, with the facilities of a sleeper bus. In fulfillment of these criteria, we present the following design:

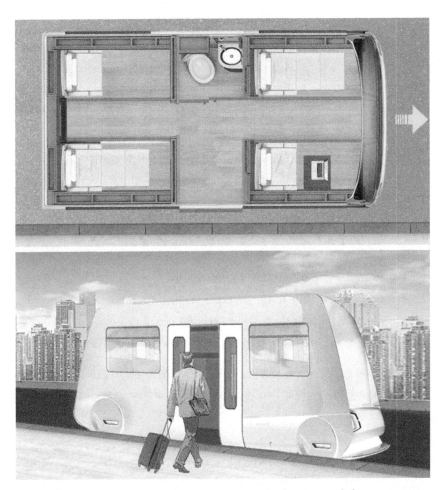

Figure 22 - This autonomous shuttle has a bathroom and four-passenger compartments.

This four-compartment autonomous shuttle features fully reclining seats, slide-out tabletops, and bathroom facilities. It can be used for overnight trips but is primarily meant to compete with short-haul air routes—of less than 500 miles.

Once unleashed on the autonomous highway of the future, these vehicles may be capable of accelerating to impressive top speeds—perhaps near 100 miles-per-hour. While this is still much slower than the 500 miles-per-hour that airliners casually attain, the tradeoff may be viable for those seeking to avoid the drudgery of air travel. Because, instead of requiring riders to hitch a ride to an airport or bus terminal, this is a door-to-door shuttle—meaning that each passenger is picked up at their residence and dropped off at their final destination address. For this reason, the number of passengers in the vehicle is kept small—so that those who are already on board won't need to wait for the car to accumulate dozens of additional passengers before starting the main stretch of the journey.

Offering door-to-door service in this fashion would eliminate much of the transactional cost of travel. The schema diminishes the burden of moving yourself (and your luggage) from:

- your home to the taxi
- from the taxi to the airport terminal
- from the airport terminal to the airplane
- from the airplane to the destination airport terminal
- from this airport terminal to a second taxi
- and, finally, from the second taxi to your final destination address.

For some routes, the passenger might actually arrive sooner by autonomous shuttle than by air—simply because no time is ever lost in such transfers.

Travel by Air

After achieving flight, airplanes are much faster than shuttles. But passengers must use other means to get to the airport. And they must contend with cramped airline seats as well as all the familiar air terminal difficulties.

Travel by Autonomous Shuttle

This autonomous shuttle is slower than an airplane. But it offers door-to-door service, much more comfortable seating, no security checkpoints, and passengers aren't required to transfer between multiple types of vehicles during the journey.

We can't know how many future passengers will prefer the serenity of an autonomous solitary cabin journey to the vivacious stimulation of air travel. While many passengers will welcome a trek that is free of transfers, some travelers truly believe that "getting there is half the fun." They may prefer the madness of crowds to the monotony of sitting in a single reclining armchair for the entire trip.

Ultimately, the top speed of the vehicle and the length of the route will be the factors that determine which form of transport will be favored for any given trip. For example, consider the LA to San Francisco route which (by car) currently takes about six hours. Flight time on this route is 1 hour and 22 minutes (with only 56 minutes spent in the air). For such journeys, the conveniences of our shuttle might not adequately compete with the speed of the flight. So the airplane might win this one.

But consider shorter routes, like:

- Los Angeles to Bakersfield: 1 Hour and 40 Minutes by car (113 Miles)
- Los Angeles to San Diego: 1 Hour and 57 Minutes by car (120 Miles)
- Los Angeles to Las Vegas: 3 Hours and 50 Minutes by car (270 Miles)

For such trips, door-to-door autonomous shuttle service might be a more attractive option. It is likely that self-driving vehicles (of one configuration or another) will someday compete with short-haul air routes and disrupt a portion of the airline industry. But America is a big place; short-haul routes comprise less than 26% of US domestic flights. Because our cities have so much land between them, self-driving cars will never be able to traverse those distances at speeds that can compete with airliners. Despite the many systemic inconveniences of air travel, it will remain the more convenient option for the majority of US routes.

In any case, the future of travel looks bright for passengers. Autonomous shuttles will provide splendid accommodations and door-to-door service for shorter hops (of under 500 miles or so). And, for longer trips, passengers will benefit greatly when airliners finally integrate their scheduling, boarding, and baggage-handling operations with the emerging networks of the coming autonomous infrastructure.

Self-driving cars will never replace airliners. But they will help to smooth the many points of friction that are currently constraining the efficient movement of people and parcels between multiple transportation platforms.

- Autonomous couriers will free passengers from the burden of luggage handling.
- Mobility as a Service (MaaS) apps will coordinate travel itineraries and payments via a single application gateway.
- And robo-taxis will facilitate each passenger transfer—as he moves from the road, to the air, to the road, and (finally) back to his home again.

Ch. 4: Upgrading Your Home's Mailbox

> The packages are...delivered by larger tubes to the city districts, and thence distributed to the houses. You may understand how quickly it is all done when I tell you that my order will probably be at home sooner than I could have carried it from here.
>
> – Looking Backward
> by Edward Bellamy (1889)

A very brief history of containerization

People have struggled with the logistical difficulties of freight transport for a long time. As maritime trade spread around the globe, ever-larger amounts of cargo had to be loaded and offloaded onto ships. In every port, stevedores were hired to pack ship hulls with freight for overseas transport. Hoisting heavy barrels and crates from ship to shore was a slow and dangerous job—that would often result in horrific injuries. To expedite this process, the shipping industry would need to develop a faster and safer way to move cargo.

Figure 23 - In the top image, stevedores load corn syrup on the Hudson River in 1912. Below, a modern container ship carries tall stacks of shipping containers. (Source: Photo by Lewis Hine.)

Wooden shipping containers were first used in 1766 by James Brindley to transport coal up the Bridgewater Canal in England. But the modern shipping container system is largely due to the 1955 collaboration of trucking company owner Malcom McLean and mechanical engineer Keith Tantlinger—who developed the "intermodal container" and "twistlock." The pairing of these two inventions enabled ships to stack tall towers of containers for transport across oceans. Once the ships arrive in port, dockside gantry cranes are used to hoist the containers off the ship and onto land—where they can be hitched to semi-trailer trucks or train cars for further ground transport.

Figure 24 - Twistlock hardware sits atop the corners of all shipping containers—allowing them to be securely stacked together for long sea voyages.

Though not much to look at, Tantlinger's patented twistlock design (and his participation in the International Organization for Standardization (ISO)), incited the creation of a globally recognized standard for shipping container construction, as well as procedures for the loading and offloading of ship cargo. This technology played a pivotal role in shaping the modern world of commerce as we know it.

The "last mile" problem

Thanks to containerization, your incoming package quite nicely flows from ship, to rail, to truck. But when the UPS guy approaches your home address, that's when the trouble starts. The most costly transportation problems usually occur in the final breadth of your package's journey—in a one-mile radius around your house called the "last mile." It is here where packages are susceptible to the greatest number of logistical inefficiencies—accounting for over 25% of total shipping costs.

So where is the rub in our rigging?

Current inefficiencies in *home delivery* are the result of many of the same problems that plagued the shipping industry in the previous century. Everything we order online (all of our groceries, granola bars, and gardening gloves) comes packaged in a different type of container. And since no single convention dictates the shape and size of these containers, delivery companies must devise clever ways to handle every possible dimension.

Additionally, the manner by which home addresses are indicated is particularly troublesome. Since there is no national code that dictates how a building or a mailbox is to be labeled, package carriers often fail to find the destination address. Also, since packages and letters may be addressed by hand, their intended destination is often illegible anyway. In 2013, the Postal Service estimated that it would cost $1.5 billion to process all of the mail that is deemed UAA (Undeliverable as Addressed) each year.

When boxes do manage to reach their target address, the occupants often aren't home to receive them. So delivery drivers must leave their cargo on the front porch—where it is susceptible to the elements as well as the prying eyes of larcenous neighbors. Around 25% of Americans claim to have suffered at least one incident of porch piracy.

Most importantly, the entire delivery enterprise requires a costly human touch. Just as the stevedores used to laboriously haul barrels onto ships (one at a time), so too must delivery drivers individually carry each package to our porches. That is to say, each delivery demands that the driver sequentially traverse the streets of your neighborhood, locate your address, park his vehicle, approach your front door on foot, ring your doorbell, and covertly tuck your package behind a potted plant. This ritual must be repeated again and again and again—for every neighborhood in the country. Each day, America's two biggest private carriers (UPS and FedEx) handle an average of

20 million packages. The US Postal Service processes 472 million pieces of mail. When pondering such staggering numbers (and the myriad of familiar logistical difficulties that delivery companies must contend with), it's a minor miracle that so many packages manage to arrive at their destination intact and on time.

Fixing the "last mile"

In recent years, there have been many noteworthy improvements to the package-delivery enterprise:

- GPS navigation helps drivers locate addresses.
- Intelligent route-planning and optimization software plots the most efficient package-delivery itinerary.
- Thanks to widespread internet availability, customers can track the progress of their packages via web interfaces.
- Electronic delivery verification, digital signatures, and onsite photos ensure that the package physically made it to its destination.
- More recently, Amazon took a mighty whack at the *last mile problem* when they released "Amazon Key" (formerly "Key-by-Amazon"). This service consists of an internet-connected camera and a mechanized door lock—which allows delivery drivers to briefly unlock the home's front door to deposit a package.

The above-listed innovations are real victories. And package delivery logistics have never been better. But many problems remain—particularly in the domain of residential delivery where *last mile failures* are still so common. The vast majority of residences are simply not equipped to accept incoming packages that are bigger than a mailbox. This is surprising given that many homes in America actually used to come equipped with a more suitable mechanism.

The Death of the Milk Delivery Door

New homeowners are sometimes surprised to discover little doors hiding in the walls of their recently purchased home. The units are easily mistaken for doggy doors or electrical utility boxes. But they were originally constructed to function as *milk delivery doors*. In the days before everyone owned a refrigerator, the local milkman made his rounds through the neighborhood—delivering milk, eggs, and groceries to every home that subscribed to the service. In some areas, houses and apartment buildings came equipped with a "milk door"—a delivery access port situated in the interior of the home's wall—usually near the kitchen. The recessed box featured two doors—one on the outside of the home and one on the inside. The milkman would deposit his delivery via the exterior door. And, once the transaction was complete, the homeowner could open the internal door to retrieve the contents.

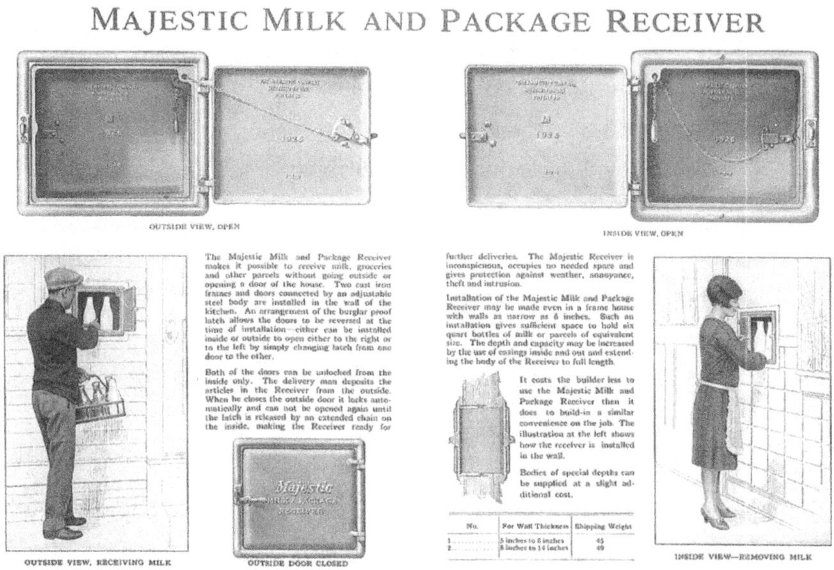

MAJESTIC MILK AND PACKAGE RECEIVER

Figure 25 - This 1920's advertisement for the "Milk and Package Receiver" (by Majestic Coal Windows of Indiana) enabled a delivery man to drop off milk, eggs, and small groceries through an exterior door. Later, the residents would access the items via the box's interior door.

In 1960, the Department of Agriculture estimated that a third of US residents subscribed to a milk delivery service. But by 2005, that number had dwindled to 0.4%. The majority of milk doors now sit unused—functioning mainly as conversation pieces for guests or as hiding places for children's toys.

Given that milk doors have been manufactured in America for over a century, you might assume that they would have evolved along with the rest of the home delivery enterprise. But, thus far, the only *shipping and receiving* conduit that can be found in most US homes is a mailbox and a porch.

Figure 26 - The only conduits by which most homes can ship and receive parcels is via the mailbox and the front porch.

These are outdated technologies. To fully capitalize upon the efficiencies of the coming age of autonomous courier vehicles, we're eventually going to need a more capable apparatus. To improve upon the process by which packages enter and exit the home, your porch will require an upgrade.

Upgrading residential shipping and receiving

Take a moment to think of your home as an organ—like a heart. It has a series of valves, veins, and arteries which all work in concert to perform a function—specifically, to pump oxygenated blood through your body. Your house has many conduits and compartments too:

- Letters enter and exit your home via the mailbox.
- Packages are placed on your front porch.
- Groceries often make their way from your car's trunk, to your kitchen counter, and then to your refrigerator or cupboard.
- Trash and recyclables leave your kitchen through a side door—ultimately to be loaded into a bin and wheeled out to the curbside once a week.
- Guests typically enter via the front door, while residents often enter via the garage—after the family car is parked.

Observe your family as they enter and exit your home during the morning rush. Take note of how much effort they must put into merely moving their bodies—from home, to school, to work, and back to home again. Note the coterie of uniformed men—who trounce up and down your walkway:

- The mailman
- The package delivery man
- The pizza man
- The trash man
- The Culligan Man

Each one of these couriers dutifully performs the elusively complex task of moving objects into and out of your home.

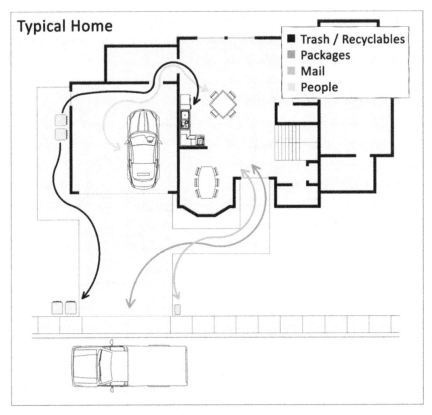

Typical Home

- ■ Trash / Recyclables
- ▨ Packages
- ▨ Mail
- ▨ People

Figure 27 - Here, we have traced the paths of people and parcels as they enter and exit a typical home.

The inefficiencies of the paths we walk in a conventional home are a result of the inefficiencies by which people and parcels are transported to the home. The contemporary floorplan of the American tract home is a result of a century of dependency on the automobile. But when *personal car ownership* is finally replaced with a ride-hailing app, we should be able to modernize the conduits by which people and parcels enter the home.

In the following image, we have redesigned the family home—arranging the rooms and entrances into a configuration that is more conducive to the age of autonomous vehicles.

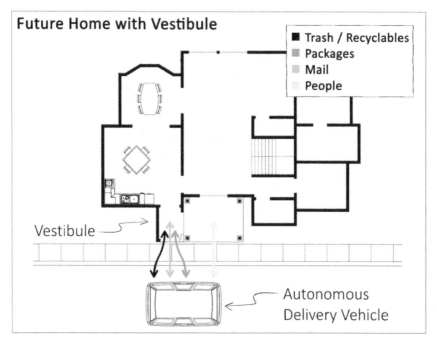

Figure 28 - Autonomous vehicles interact with a mechanized dropbox in this home's street-facing vestibule.

First, note the absence of a garage and a driveway. Since nobody will own a car in the future, our house requires neither. There is no setback between the main structure and the sidewalk. Instead, the building is situated near the street—making it readily accessible to autonomous delivery vehicles. Most importantly, all incoming packages are ported through one structure—a mechanized dropbox embedded inside the home's vestibule.

Traditionally, the vestibule may refer to any small room or antechamber that protrudes from a structure. It often acts as an entryway and sits adjacent to the front door. Sometimes a vestibule will function as a security checkpoint, a coat room, or a mudroom for homes in colder climates.

Figure 29 - Traditionally, a vestibule is an antechamber, hall, or lobby that extends the main entrance of a building.

From a distance, our future vestibule might look similar to a contemporary one. But it will contain hardware that enables it to function as a delivery conduit and a beacon—used by autonomous vehicles to locate your residence and rapidly offload cargo.

Given the rapid growth of e-commerce, we will soon need to find a better way for our homes to interact with the autonomous couriers of the future. We'll need a more efficient mail delivery system—one that facilitates the residential-delivery process. In the following image, we have outlined how such a system might work.

Home Vestibule with Mechanized Dropboxes

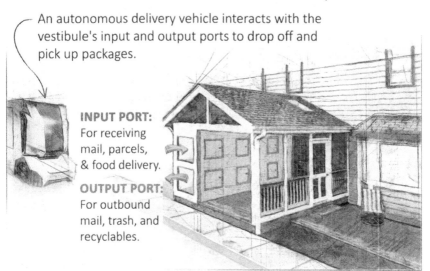

An autonomous delivery vehicle interacts with the vestibule's input and output ports to drop off and pick up packages.

INPUT PORT: For receiving mail, parcels, & food delivery.

OUTPUT PORT: For outbound mail, trash, and recyclables.

Figure 30 - The vestibule has an input and an output port—through which parcels can enter and exit the home. When the autonomous delivery truck approaches, the input door will open—allowing the truck to deposit a package. When the family needs to ship something, the outgoing package is placed into the vestibule's output port and a pick-up is requested online.

Our "vestibule of the future" is basically an internet-connected dropbox that acts as the home's "shipping and receiving" apparatus. Note that, in our above diagram, we have exaggerated the size of the ports as our intent is not to engineer the exact mechanics by which deliveries of the future will be received. Instead, we wish to convey that future autonomous delivery vehicles will need *some* type of mechanism into which packages can be rapidly deposited. Ideally, architects will someday succeed in thoroughly integrating such devices into future homes. For example, in the following rendering you'll notice that a street level dropbox has been nicely incorporated into a traditional chateau design.

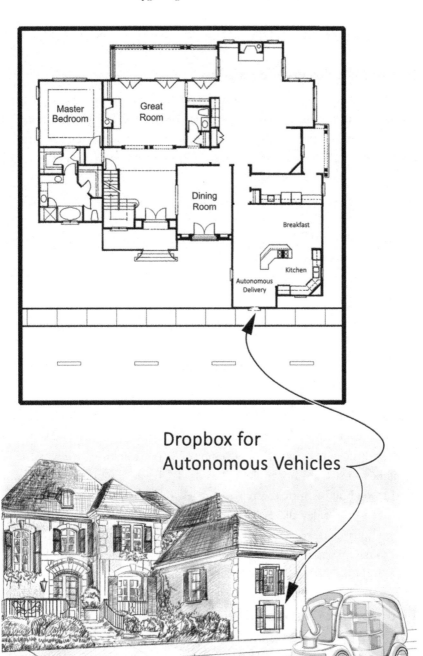

Figure 31 - A portion of this home has no setback; it sits adjacent to the sidewalk and a mechanized dropbox (embedded in a faux window) can receive deliveries from incoming autonomous couriers.

We'll take a moment to list four of the main features of such devices now.

Feature #1: Street-Accessibility

The most challenging leg of a parcel's journey is not in the crossing of the Pacific Ocean nor in the costly "last mile" of home delivery. Instead, it's in the "The Final 50 Feet"—that gap of terrain that exists between the delivery truck (sitting idle at your curb) and the home's entryway. Many types of obstructions lie along this path:

- picket fences
- security gates
- barking dogs
- garden sprinklers
- bridges
- multi-step porches
- steep driveways
- and (the biggest obstruction of all) a locked front door—when no one is home to receive the package

Traversing this obstacle course is one of the most costly and challenging problems for delivery companies. Given that such impediments make home delivery difficult for trained human employees, one can imagine how difficult it will be to teach a robot to do it. Thankfully, because future homes do not need to be built with driveways or garages, our vestibule's dropbox can sit adjacent to the sidewalk. By doing away with lengthy residential setbacks, autonomous delivery vehicles can rapidly offload their cargo without the need to navigate across the logistical minefield that exists between the street and your home's front door.

Feature #2: Mechanized and Secure

When a drop-off is underway, an appendage will extend out from the autonomous delivery vehicle, and enter the vestibule's input port. The port door will then open to allow for the depositing of packages. When the drop-off is complete, the port door will immediately close.

Unlike your mailbox, the vestibule's port door can only be opened by designated delivery vehicles. Hence, the contents of the vestibule will only be accessible to the homeowners themselves—not their neighbors. Such a system should eliminate most types of porch piracy.

Figure 32 - An autonomous courier makes a deposit into this home's mechanized dropbox. Because the dropbox's port is situated near the street, the delivery vehicle can complete the exchange in just a few seconds.

Feature #3: Internet-connectivity

Like every other future appliance, the vestibule's hardware will be connected to the internet at all times. As packages flow in and out of its ports, residents will receive notifications about the vestibule's state.

- Did your delivery of dishwasher soap arrive yet?
- Has the courier picked up your outgoing package of Christmas cookies for Aunt Edna?
- Is the dropbox currently full of packages, and needs to be emptied?
- Is the port door locked or in need of repair?

The dropbox in the home's vestibule will be constantly on-call—communicating with the residents via mobile app—just as if it were their own 24-hour doorman.

Feature #4: Geo-locatability

Currently, delivery drivers must manage all of the familiar complexities of vehicle navigation along with the many confusing road signs, inconspicuous driveways, and obstructed markings that lead to the destination address.

Figure 33 - To locate your home, delivery drivers must decipher road signs and address plaques with varying degrees of visibility and legibility. (Photo: Famartin)

The dropbox hardware in our vestibule has its own IP address as well as its own associated geo coordinates. So instead of relying on wall-hung address plaques or destination approximations, autonomous courier vehicles will have access to the exact latitude and longitude of the vestibule itself. Thus removing any ambiguity about the location of the package's final destination.

Curb-to-doorstep delivery for existing homes

The vestibule dropbox design that we have showcased above is for a new housing development. Existing tract homes often have a setback of ten to twenty feet. They can't be easily relocated such that they butt up against the street. Instead, a more typical retrofit is depicted in the following rendering:

Figure 34 - A single-family home has been retrofitted to accommodate autonomous delivery vehicles.

Let's list three of the changes we've made to the home in this rendering:

1. Thanks to the end of personal car ownership, garages and driveways are no longer needed. So the garage has been converted into a living room and the home's original bay window design has been replicated on its facing wall.

2. The asphalt driveway has been removed and landscaped over. A new walkway extends from the street to the home's front porch.

3. Most importantly, a street-accessible dropbox has been installed near the curb—in which autonomous delivery vehicles will deposit packages via a mechanized exchange. The vast majority of home deliveries occupy less than one cubic foot of space. So most incoming parcels will fit in a modestly sized dropbox of the sort depicted here.

Such modifications are representative of the types of residential retrofitting that will be common during the transition to an autonomous infrastructure. This job was easy—because it's a one-story house without any steps or steep grading. However, while the vast majority of Americans live in such single-family detached homes, about a quarter of us live in multi-story apartment buildings. Stairs are not a robot's friend. But each residence has its own unique set of obstacles that lie between the street and the front door—e.g. elevators, ponds, fences, bridges, or rock paths. It will be tricky to design an autonomous delivery vehicle that is capable of circumnavigating *every* type of obstacle. But this challenge hasn't stopped tech companies from filing patent claims.

In what is perhaps the most entertaining patent ever ratified (US 10,514,690 B1), Amazon's engineers sketched out a dozen methods by which future autonomous couriers and drones might cooperate to deliver your latest e-commerce order of biscuits and birdseed. Some of their figures are printed on the following page.

Patent No.: **US 10,514,690 B1**
Date of Patent: **Dec. 24, 2019**

COOPERATIVE AUTONOMOUS AERIAL
AND GROUND VEHICLES FOR ITEM
DELIVERY

Applicant: **Amazon Technologies, Inc.**

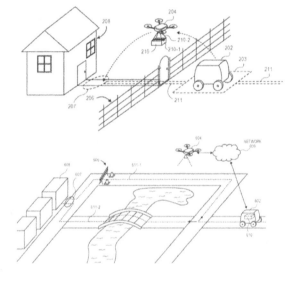

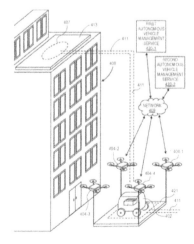

Perhaps, for some products, home deliveries will be completed utilizing a combination of drone and AV technology. But curb-to-doorstep delivery may not be possible for every type of home—at least not in the early days of our conversion to an autonomous economy. Instead, some residents may have to retrieve their incoming packages from a designated meeting place or a central neighborhood distribution hub—similar to an Amazon Locker. This more down-to-earth approach is reflected in Amazon patent US-20180024554-A1 which describes a delivery process in which multiple AVs rendezvous with a delivery truck at a central neighborhood meeting location.

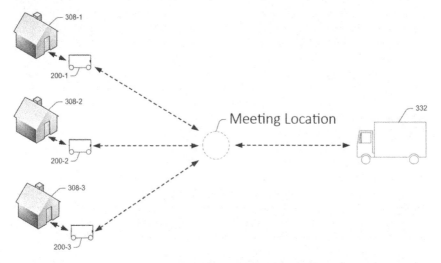

Figure 35 - Amazon patent (US20180024554A1) describes a procedure for delivering packages to a "meeting location" at which local residents or smaller autonomous couriers can rendezvous to receive packages.

An alternate solution to the *The Final 50 Feet* problem comes from the Oregon-based tech company Agility Robotics—founded in 2015 by Jonathan Hurst after researching bipedal motion for his Ph.D. thesis at Carnegie Mellon University. The company's prototype (codenamed "Digit") can carry a 40-pound box across different types of terrain and even upstairs.

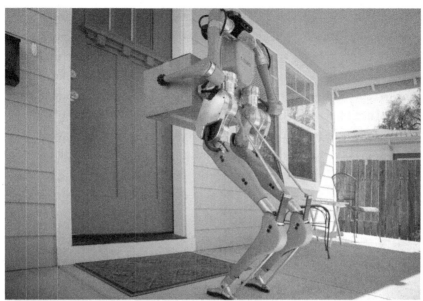

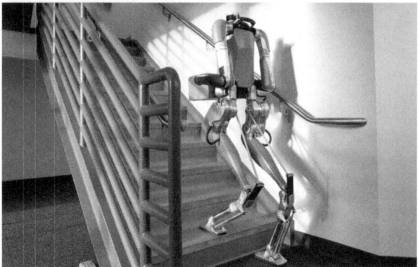

Figure 36 - The Oregon-based tech company Agility Robotics hopes to couple their biped robots with autonomous delivery vehicles.

Partnering with Ford, Hurst plans to piggyback his robots onto their autonomous delivery truck prototypes. In theory, future delivery trucks could park in the middle of a neighborhood, and a squad of two-legged bots could hop out and sprint deliveries to every porch in sight.

Performing such a task is no small feat. It will probably take engineers (like Hurst) another couple decades to teach their robot how to unlock a fence gate, tread carefully across a garden path, deter a barking dog, and leap up a concrete stoop—all with the same grace as a uniformed UPS man. However, given the impressive physical feats that his bots are already capable of performing, there is little doubt that they will eventually acquire the finesse needed to complete such actions. And thus, we can casually infer that (at least someday) *all* "delivery men" will be robots.

In listening to Hurst's talk at the Carnegie Mellon 2016 Robotics Institute Seminar, one gets the impression that he is quite aware of his invention's potential. During concluding remarks, Hurst was forthcoming about his future goals:

I want to change the retail economy... Imagine...a fleet of autonomous [trucks and robots doing] curb-to-doorstep delivery. [Package deliveries would] cost ten times less... [If we could do that then] why would you shop? [There wouldn't be much of a] reason to go to stores... [Instead, you'd] just [buy something on] Amazon...and [it] shows up on your doorstep within half an hour...

The new trade winds

Making *automated product delivery* available to every home in America may require the marriage of self-driving courier vehicles and other types of automatons. Future homes will be equipped with an appliance that will interact with these machines to receive incoming packages. This device may perhaps come in the form of a mechanized dropbox (embedded in a home's street-facing vestibule), an electronic front door lock (like "Amazon Key"), or a neighborhood delivery hub. While we can't know how these systems will evolve, we do know that—given the rising volume of e-commerce deliveries—there will come a day when order fulfillment does not require *any* human participation whatsoever.

When such a hands-free delivery infrastructure is finally developed, newly purchased products will be tossed from warehouse robot, to wheeled delivery robot, to courier robot, and (finally) into the mechanized dropbox at the destination address—all without any human involvement in the transaction. Future consumers will quickly grow accustomed to near-instantaneous delivery. From their perspective, the product will simply substantiate in their living room—perhaps just thirty minutes after they buy it online.

The implications of such a rapid fulfillment schema cannot be overstated. We're not just talking about a better way for eBay sellers to ship popcorn poppers and printer ink. Instead, the utilization of autonomous courier vehicles will torque the wheels of commerce, just as the advent of shipping containers hastened the winds of international trade. Given that business is so inextricably bound to the speed by which services are rendered and goods move down a supply chain, increasing the deliverability of each component part—as it flows from manufacturer to consumer—will have obvious positive economic repercussions.

When these things come to pass (when every home in America features a *retail delivery receptacle* and every product is hauled on the backs of subservient automatons), national markets will achieve new heights of production, competition, variety, affordability, and trading volume—inciting a frenzy of economic activity that has never been seen before. That's good news for consumers. The bad news is this: two of the most common jobs in America—delivery driver and retail salesperson—will no longer exist.

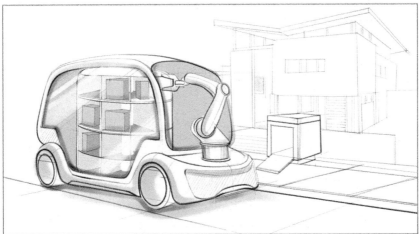

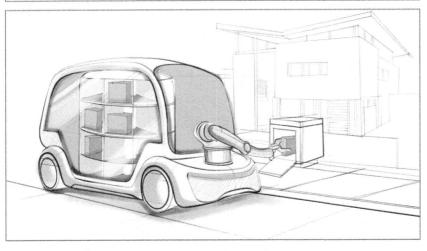

Ch. 5: How AVs will Change the Meal Delivery Industry

Food in a Pill

Right up there with flying cars and moon colonies, "food pills" are another classic sci-fi trope that failed to materialize in the 21st century. The idea that one could consume an entire chicken dinner in a single pill—without the laborious chore of cooking or chewing—seems to have tickled the fancy of both futurists and tired housewives alike. Prior to the Chicago World's Fair of 1893, the political activist Mary Elizabeth Lease was asked to predict what inventions the world of tomorrow would bring. She wrote:

A small [vial]…will furnish men with substance for days. And thus the problems of cooks and cooking will be solved.

Given that she was a firebrand for Women's suffrage, it was probably easy for Mary Lease to fantasize about a world where women could serve their family an entire meal in a "small vial"—thus freeing themselves from the yoke of daily kitchen duty.

Present-day corporations are still trying to tap into this desire. Vitamin and supplement manufacturers, weight loss programs, and tech startups all offer meal-replacement solutions—in the form of protein bars, energy drinks, and shakes.

Most famously, the Los Angeles-based startup *Soylent* has become popular with time-conscious millennials. Launched in 2013 by software engineer Rob Rhinehar, Soylent comes in the form of a 400-calorie vanilla-flavored shake.

It purportedly contains all the nutrients of a well-balanced meal but at a fraction of the cost. Upon conceiving of his invention, Rhinehar insists that he reduced his food bill from $470 to $155 per month.

Given that the US dietary supplement industry is valued at over $122 billion, it's clear that many of us have not yet given up on the dream of consuming a *meal in a pill*. But, while almost all of us have sampled such supplements, most of us still prefer to eat our food on a plate. This predisposition is due to a myriad of psychological and neurophysiological reasons:

- First, meal-replacements are typically consumed in seconds— without the social ritual of communal eating nor the tactile feedback of crunching and chewing. Hence, they often fail to arouse sensations of satisfaction.

- Additionally, mammals are simply not meant to eat the same food again and again. Via a phenomenon called "sensory-specific satiety," our brains depreciate the taste of a food type if it is consumed too often. Since most of us don't have the willpower to override this circuit for very long, most of us quickly tire of a diet consisting of protein shakes and PowerBars.

- Finally, it's possible that the transmogrification of food (via cooking) played a vital role in altering our evolutionary makeup. In his book "Catching Fire," Harvard anthropologist Richard Wrangham argued that the practice of cooking yielded a more energy-dense food product which allowed for the evolution of more sophisticated brains. It could be that cooking was even more instrumental in shaping the course of human evolution than language or tool-making. So attempting to bypass the dining ritual by swallowing a couple protein pills may forever fail to appease the hungry savage that resides within us all.

To the chagrin of retro sci-fi lovers, we will probably never be able to divorce the ritual of eating from the human experience. Even if we could squeeze 2,000 calories into a single pill, our hunger would never be properly satiated and such diets would have little lasting appeal. Instead, we'll need to find *some* way to make the chore of preparing hot food more manageable.

Nobody wants to cook

In the life of the average American, he'll spend about 26 years sleeping, 4 years driving, and 5 years consuming glorious food—including meal preparation, shopping, eating, and cleanup. We can't do much about the 26-year sleeping requirement and we've already talked about reducing the 4 years spent behind the wheel. So now, let's try to tackle the third item in our list—let's describe how autonomous vehicle technology might help to reduce the amount of time we spend preparing meals.

As evidenced by the international success of The Food Network, the billion-dollar cookbook publishing industry, the dozens of reality TV cooking shows, and the bottomless number of foodie videos on YouTube, it's clear that many of us enjoy the *spectacle* of food preparation. But, when it comes to actually preparing food for ourselves, we're not too jazzed about it. Americans are in love with *the idea* of cooking much more than they love cooking itself.

In 2018, the US Bureau of Labor Statistics reported that the average household spent $3,459 on takeout food annually. A 2013 study by the National Institutes of Health found that the number of meals consumed at home by middle-income households dropped from 92% (in the mid-1960s) to 69% (by the late 2000s). The pitch of this drop varies slightly—depending on the income and age of the respondents. But the amount of time spent on food preparation has fallen for *every* demographic, at *every* level of income.

It might be tempting to blame these trends on the socioeconomic disruptions of the 1960s—like the entrance of women into the workplace or the advent of second-wave feminism—which awakened women to the Feminine Mystique and implored them to recognize the existential harm of baking cookies for the patriarchy. Such societal shifts undoubtedly played a role in changing our eating habits. But, while it is true that working women do cook less than women who don't have an outside job, the difference is minor. Women without jobs don't like to cook either—they spend only 20 minutes more in the kitchen than their workaday counterparts.

In 2017, former Cambridge Group director Eddie Yoon conducted a survey asking Americans to describe their feelings about the task of cooking. Only

10% of consumers said they "love to cook." 45% were lukewarm about it. And 45% said they "hated it." This leaves us with the unsurprising result that—when given the choice to cook or *not* to cook—90% of us would rather not cook anything at all.

So where are these people getting their meals then?

Where Americans get their meals

We get most of our "ready-to-eat" or "ready-to-heat" meals from five sources:

1. The fast-food drive-through window
2. The restaurant take-out counter
3. Delivery from the local pizza guy
4. Delivery via app-based delivery services—like UberEATS, DoorDash, and GrubHub
5. Frozen microwaveable "TV Dinners"—purchased from your grocer's freezer

How we get our ready-made meals

| Frozen microwavable "TV Dinners" | Restaurant Take-out | Restaurant Delivery (Often via Uber Eats, DoorDash, or GrubHub) |

Pizza Delivery

Fast Food
Drive-through

We can organize these five sources into two groups:

1. Frozen food—typically in the form of microwavable dinners.
2. Takeout food—meals that are prepared in a restaurant kitchen and purchased at the drive-through window, the take-out counter, or (more recently) via app-based delivery services.

Let's discuss the pros and cons of both of these meal sources now.

Meal Source #1: Frozen "TV Dinners"

While packaged frozen meals have been sold for over a century, the first company to mass-produce a nationally successful product was Swanson. The Swanson family has been in the food business since 1899. But it wasn't until 1951 when the Swanson brothers divided an aluminum tray into three compartments (for turkey, mashed potatoes, and peas) and successfully marketed their meal-in-a-box solution to busy housewives. Since half of US households owned a television set in the mid-1950s, the new Swanson frozen meals were branded as "TV Dinners," thus further solidifying their market positioning as a "convenience food"—able to be prepared quickly and consumed in front of the television set.

Figure 37 - This 1954 ad for Swanson TV Dinners features a family eating three Swanson meals in front of a television set.

Swanson TV Dinners were an immediate hit with the time-strapped public. They remain so to this day; each year, the average American consumes 72 frozen meals (1 in 5 dinners). Despite recent declines in television viewership, the initial Swanson branding ("TV Dinners") has managed to stick around— still used to represent *any* frozen meal on offer in your grocer's freezer. According to the American Frozen Food Institute, frozen meals are now a $56.7 billion industry in the US alone. A 2014 market paper by Persistence Market Research reported that the global frozen food market would achieve a value of $156.4 billion by the end of 2020.

Given such high dollar amounts, it appears that many of us are electing to prepare our meals via microwave, rather than cook them from scratch.

So what's wrong with that? What's so bad about frozen TV Dinners?

Below, we've listed three of their main faults.

- Most people are quick to note that TV Dinners usually don't taste as good as a fresh "homemade" meal. During the freezing process, ice crystals form which ruins the texture, taste, and color of the food. The thawing and heating process itself is an error-prone procedure. Microwaves often fail to heat each portion of the meal at the same temperature—resulting in cold spots or over-cooked portions.

- Second, not all types of food can be placed in a frozen meal box. You won't find any salad or bread rolls in the side-tray of your *Hungry Man TV Dinner*. Some foods tolerate the freezing and thawing process better than others. Consequently, the selection in your supermarket's frozen dinner section will forever be limited to those foods which are capable of resembling their original form after microwaving.

- Finally, most TV Dinners are not very good for you. Contrary to popular belief, it isn't the freezing process itself that makes the meals "unhealthy." Nor is the microwave radiation to blame. Instead, the types of frozen meals that Americans buy tend to be of the more unhealthy variety; there's a qualitative dietary difference between a bag of frozen vegetables and a box of frozen pizza pockets. Such meals are heavily processed, high in trans fats, calories, sodium, and sugar. They often come with preservatives such as butylated hydroxytoluene (BHT) and they may contain partially hydrogenated vegetable oil.

The unhealthy state of our frozen dinners is typically not very surprising to anyone. As we toss boxes of microwaveable Salisbury Steak into our supermarket carts, most of us are quite aware that we haven't made the wisest meal choice. Whatever multitudinal factors are at play in our nation's health crisis and obesity epidemic, it is certain that convenience foods are in part to blame. Contrary to the advertisement messages conveyed in the glossy

Swanson magazine ads of the 1950s, frozen TV dinners will *not* become the meal-replacement product of the future.

Meal Source #2: Takeout Food

"Take-out food" is the most common method by which Americans attain ready-to-eat meals. These days, they are sometimes brought to us by delivery drivers (from DoorDash, GrubHub, or UberEats). But most commonly, we get them from drive-through windows—which (for some franchises) account for an astonishing 60% to 70% of food sales. It's not uncommon to see the drive-through window of your local McDonald's (or Burger King, or Taco Bell, or Wendy's) inundated by a serpentine queue of cars—each one manned by an irritated driver—waiting to satisfy his hunger with a supersized meal deal. The ancient ritual of eating has devolved into an unfortunate circus routine—in which a corporation tosses bags of burgers through a car window, and the recipient executes the perilous stunt of *eating while driving.*

Even more distressing is the frequency with which this trick is performed. As Cheryl Fryar (lead author of the CDC's 2018 Health and Nutrition Examination Survey) told CNN, "On any given day in the United States, an estimated 36.6% [of adults]…consume fast-food…" The stats are even worse for younger adults (ages 20 to 39)—a distressing 44% of them consume fast-food daily. If trends continue, it will soon be the case that half of the country will be dining at a fast-food franchise location every single day.

Fast-food is not very healthy

All of us know that the American fast-food diet does not lend itself to the healthiest lifestyle choice. Fast-food is high in calories, saturated and trans fats, sodium, and sugar. When consumed in excess, fast-food is commonly considered to be a major contributor to the *national obesity epidemic*—which is linked to Type 2 Diabetes, cardiovascular disease, cancer, and perhaps responsible for up to 18 percent of US deaths (for citizens age 40 to 85). It's clear that the country is paying a dear price for our addiction to foods of convenience.

Curiously, while US industry has managed to come up with many modern-day conveniences, the chore of simply obtaining a healthy meal remains a costly endeavor. Drive-through windows are still the fastest and cheapest way to get a quick bite. Pizza delivery is the second most common option. But, oddly, the evolution of food-delivery seems to have stopped there. Obtaining a prepared hot meal—one that is simultaneously fast, cheap, and healthy—has remained an elusive goal for decades.

The promise of "Home Delivery" and the Dot-Com Bubble

Internet CEOs have long heralded the eventual arrival of effortless online home delivery. Readers who are old enough to remember the first dot-com bubble might also recall its two most famous causalities—Kozmo and WebVan.

- **Kozmo** was known for its legion of dedicated bike messengers— each one festooned with a branded helmet and a garish orange messenger bag. Kozmo founder Joseph Park envisioned a day when smaller items (like books, DVDs, meals, and home essentials) could be pedaled to the doors of online customers—eager to pay for the convenience of home delivery. At their peak, Kozmo's uniformed couriers rolled down the streets of eleven major US cities—from New York to San Diego. Shortly after its founding in 1998, the company raised $250 million in seed funding. Three years later, when the dot-com bubble burst, they laid off each of their 1,100 employees and closed their doors for good in April of 2001.
- **WebVan** was valued at $1.2 billion and owned massive computer-controlled warehouses—designed to load customer groceries into fleets of brightly colored delivery trucks. After only three years of operation, WebVan had accumulated a deficit of over $830 million. They too filed for Chapter 11 bankruptcy after the 2001 dot-com apocalypse.

Home Delivery 2.0

A lot has happened since the first dot-com bubble burst. Technology has indeed improved—specifically, it's all gone mobile. The ubiquity of internet-enabled cellphones and the advent of app-based employment has allowed for a new type of *courier class*. Anyone with a phone and a car can get work as a delivery or taxi driver—ready to transport parcels and people via the emerging gig economy. Mobile phones have enabled meal-delivery companies to manage the logistical nightmare of finding a human that is willing to drive a pastrami sandwich over to your house at 9:30 p.m. on a Tuesday.

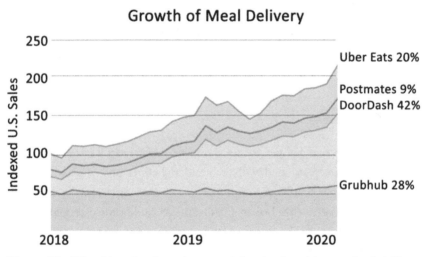

Figure 38 - Monthly sales have increased for the four biggest food delivery companies: DoorDash, GrubHub, UberEats, and Postmates. (Plot by SecondMeasure.com.)

As evidenced by the above chart, app-based meal delivery is finally starting to catch on. Allied Market Research has estimated that the global food delivery market could hit $16.6 billion by 2023. As things look now, the three largest food delivery companies—UberEats, Postmates, and DoorDash—are all steadily growing. But the road from the restaurant to your front door is a windy one.

Currently, meal-delivery websites are plagued by all of the same *last mile problems* that every other delivery company must suffer through. Their challenge is further complexified by the time constraints of their cargo—a hot meal doesn't stay hot forever. Ready-to-eat meals only retain their ready-to-eat status for a narrow window of time. As the minutes tick by, steak loses its sizzle, soup cools, and ice cream melts. Such difficulties were epitomized in a 2015 lawsuit between DoorDash and the California burger chain In-N-Out. Here, In-N-Out claimed that they had:

...no control over the time it takes [DoorDash] to deliver [In-N-Out]'s goods to consumers, [nor] over the temperature at which the goods are kept during delivery, nor over the food handling and safety practices of [DoorDash]'s delivery drivers...

Known for their commitment to using high-quality ingredients, In-N-Out management was furious to learn that some of their customers were receiving deliveries of soggy burgers and wilted french-fries. In-N-Out menu items are meant to be eaten as soon as they leave the chef's hand. But DoorDash drivers don't always manage to fulfill their orders in the timeliest fashion.

As the case was being litigated, many legal commentators and home delivery companies chimed in with their own opinions and arguments—which made for some amusing internet drama in 2015. For a while, it seemed like the entire home delivery business model might be in jeopardy. But, anticlimactically, DoorDash and In-N-Out ultimately arrived at an undisclosed settlement, and DoorDash quietly removed the burger franchise from their online menu.

The grievances cited by In-N-Out are a product of the systemic inefficiencies of home delivery. Consider the many potential fault points in the process:

1. First, when an order comes in, delivery drivers must navigate to the requested restaurant—which may reside in an area of town that is unfamiliar to them.
2. Then, if the establishment doesn't have a drive-through window, the delivery driver must find a parking spot, get out of his car, and retrieve the customer's meal from inside the restaurant.
3. Then he must get back into his car, drive to your address, locate your apartment door, and hope that someone is home to receive the goods.

All of these steps must happen within a time frame that is short enough to prevent your food from achieving room-temperature. And each step must be completed at a price that allows the delivery company to turn a profit. The fee for this service is paid for by both the customer and the restaurant—which is often charged a commission of around 20% per order. The driver gets to keep the tip and receives a base pay for each delivery—which (depending on the app) may amount to a mere pittance.

In light of the above, it's easy to see why so many home delivery startups have come and gone in the last two decades. Finding someone to pick up a chicken-fried steak at your local diner, drive it to your neighborhood, find your address, park their car, knock on your door, and handoff your food (with a smile) is a logistically tedious and expensive undertaking. The many points of friction in this schema inflate the meal price—making it prohibitively high for most Americans to use on the daily.

If we are to improve upon the national diet (and get people to stop eating fast-food), we will first need to improve upon the efficiency by which nutritious meals are delivered. This entails upgrading our current delivery infrastructure—creating a new system that utilizes the many transportation innovations that future autonomous vehicles will be capable of.

Dark Kitchens + Autonomous Meal Delivery

A brief history of Dark Kitchens

Following the advent of meal delivery apps, many restaurateurs were surprised to receive new customers from the emerging online marketplace. For some eateries, sales from UberEATS and DoorDash began to outpace their brick-and-mortar receipts. Consequently, some owners decided that they no longer needed to bother with the complexities of running a traditional restaurant. So they locked their front doors, pulled the shades, and closed their dining rooms—thus making the restaurant appear "dark."

A "dark kitchen" (also called a ghost kitchen, virtual kitchen, virtual restaurant, or cloud kitchen) is a restaurant that only exists online. This sort of business model is only possible because, when a customer orders a meal from a restaurant in cyberspace, he can't see any of the traditional restaurant selling points—like location, dining room ambiance, view, customer service, or décor. From the online user's perspective, he just sees a webpage with a menu and a restaurant logo. It is possible that this logo represents an actual brick-and-mortar restaurant location. But these days, it's becoming increasingly likely that the logo is just one of a dozen brands run by the same owner—who might be operating each dark kitchen out of single commercial food-prep facility.

While some dark kitchens have a large staff, others are small-time operations—possibly consisting of just one employee—the cook. The restaurant itself may not even have a physical address. Meals can be prepared in a commissary (or shared-use) kitchen which may be rented on a monthly or even an hourly basis. Such facilities may function as "kitchen incubators" — offering services to new restaurateurs like refrigeration, storage, food-prep workstations, order-processing software, and even entrepreneurial services like marketing and accounting.

Recently, many investors have been trying to cash in on the dark kitchen craze. This includes independent restaurateurs, food delivery companies (GrubHub and UberEats have both invested millions), and even tech companies like Google Ventures—who, in October of 2018, announced the closing of a $10 million Series-A funding round with "Kitchen United."

At Kitchen United's Pasadena branch, six independent restaurants reside in one shared-use kitchen. There are no dining rooms or waitresses at Kitchen United of course. All of their six branded restaurants only exist online.

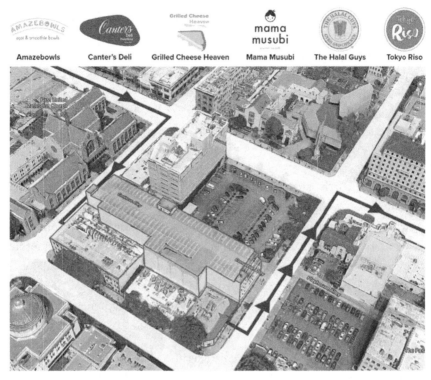

Figure 39 - Six independent restaurants share one kitchen facility on the ground floor of Kitchen United's Pasadena location. Delivery drivers enter from the west parking garage entrance, pick up their orders, and then proceed on to the destination address.

Kitchen United calls these buildings "kitchen centers." Given the success of their Pasadena location, they plan to open fifteen additional kitchen centers over the next several years.

Kitchen Centers of the Future

The emergence of such kitchen conglomerates and the recent widespread use of online ordering apps are a result of the transitional state of the conventional restaurant business model. Facilities like Kitchen United offer us a glimpse into the future of meal preparation and delivery. Currently, their drivers must circumnavigate all of the typical delivery problems that we have discussed throughout this book. But a day will come when deliveries will be expedited by autonomous couriers. Such vehicles will retrieve newly cooked entrées from meal preparation facilities (like Kitchen United Pasadena branch), race the meals across town, and drop them off at the addresses of the hungry consumers.

Let's take a moment to describe how this process might work.

Step 1: Prepare the customer's meal order

The way that we'll order online meals in the future will not be dissimilar to the present method. Customers will pull out their cellphones, view an array of available dishes, select their favorite entrée, and purchase the item via an associated payment method.

It's possible that this order will flow into the hands of the mom & pop diner around the corner. But as online meal delivery continues to catch on, larger kitchen facilities will be needed. In the following image, we have sketched a high-volume meal-production facility—with four independently operating chain restaurants.

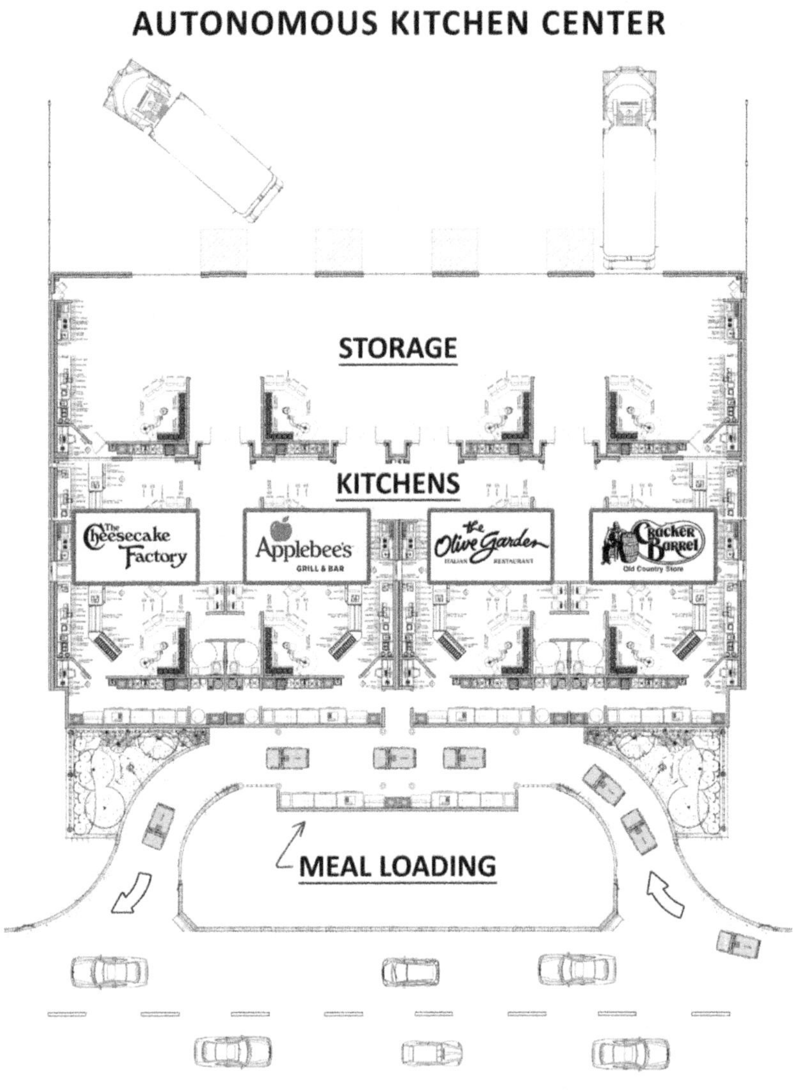

Figure 40 - Four restaurant kitchens occupy this large "Autonomous Kitchen Center." Autonomous vehicles approach from the right, retrieve the customer's order, and immediately exit to make their delivery.

No customers are allowed in these restaurants. All orders arrive via internet and exit the facility on the backs of autonomous delivery vehicles. This location's design would be appropriate for a major city—where millions of single-serving meals are produced and consumed each day. Given the speed and efficiency of the coming autonomous infrastructure, such facilities should be capable of serving thousands of people in a radius of perhaps thirty to forty miles.

As individual meal orders come into the kitchens, chefs, line cooks, and food packers will work in unison to prepare the meal and get it ready for transport—which means they'll need a container to put it in.

Step 2: Place the meal in a delivery container

Currently, as take-out orders come in, every restaurant performs the same function—they prepare the food and place it in a container for pick-up. But since there is no nationally recognized standard for food delivery packaging, each restaurant is using a different type of container.

Figure 41 - "Takeout food" is packaged in many different types of boxes and bags.

To reduce the inefficiencies of residential meal delivery, the restaurant industry will need to eventually agree upon a containerization standard. This is especially pressing if they hope to load various types of meals into the cargo hold of a 3rd party autonomous delivery vehicle. Meal package standardization will be necessary to avoid vehicle loading and unloading problems, and to allow for the vehicle's software to appropriate cargo space for multiple containers.

Here, we present a prototype design for a single-portion meal container. Like a *bento box*, the container features partitioned compartments for holding different food types.

Delivery Box with Hot Meal and Beverage

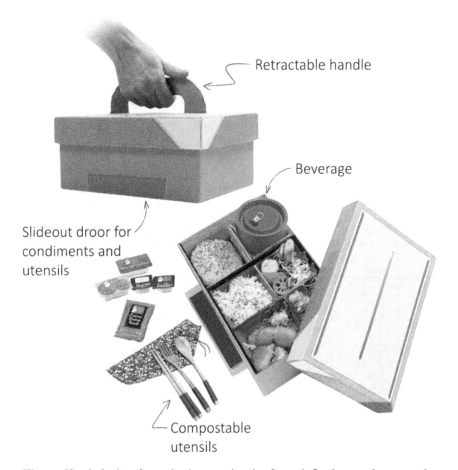

Retractable handle

Beverage

Slideout droor for condiments and utensils

Compostable utensils

Figure 42 - A design for a single containerized meal. Such containers can be easily loaded and stacked into the cargo hold of autonomous courier vehicles.

Given the many shapes and sizes that our food comes in, containerizing home delivery will be a tricky task. Pizzas are flat and round, cakes are cylindrical, tacos are skinny, and a bucket of chicken is shaped (well) like a bucket. Undoubtedly, many different container configurations will be utilized. But so long as some stackable convention is agreed upon, future autonomous couriers should be able to properly store the cargo and keep it secure enough to survive the speedy trip to your front porch.

Step 3: Deliver the meal

After the meal has been containerized and loaded into the delivery vehicle's cargo hold, it's time to start the journey to the customer's address. The same vehicle may be loaded with multiple meals from multiple different kitchen centers. In our diagram here, we suppose that five customers have requested a meal order—three from Kitchen Center A, and two from Kitchen Center B. Our courier vehicle visits both kitchen centers to retrieve each order and then proceeds to the residences to drop off its cargo at each house.

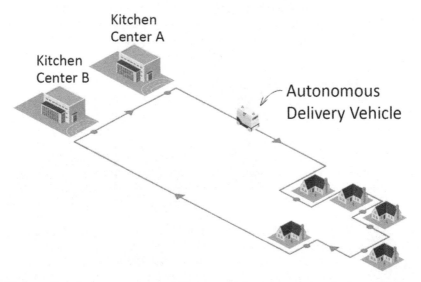

Figure 43 - An autonomous delivery vehicle retrieves meals from Kitchen Center A and B, and then proceeds to drop off its cargo to each of the five nearby residences.

The vehicle's navigation software will select the ideal route for pick-ups and drop-offs—continuously optimizing this route as existing orders are fulfilled and as new orders come in. Future delivery companies will own legions of such vehicles—each one will buzz up and down the streets of suburbia— delivering burgers and burritos throughout the day and night.

Step 4: Drop off the meal

In the previous chapter, we described your future home's "vestibule"—a street-accessible dropbox that is capable of receiving packages from autonomous delivery vehicles. While we don't know exactly how this exchange will work, it is clear that future couriers will need some sort of mechanized apparatus in which to offload their cargo. In the below renderings, autonomous couriers are seen making deliveries to multiple dropzones.

Figure 44 - Autonomous couriers will be able to deliver meals to many different types of locations.

You'll notice that order deliveries need not be limited to residences. Autonomous couriers will be capable of delivering meals to many different types of locales—including parks, entertainment centers, or even to other restaurants and bars. In the future, some public venues may not directly monetize their facility via food sales. Instead, they may be specifically designed to accommodate incoming third-party meal deliveries and might encourage their customers to place such orders. (We'll be talking about such hybrid restaurants in Chapter 6.)

Step 5: Discard the waste

Finally, after the meal has been consumed the waste must be disposed of. This final step of home delivery should not be overlooked. Because, as more single-portion meals are purchased online, more containers will be needed to transport them. If every American (all 328 million of us) were to start eating dinner in the above-described fashion, then we would rapidly be filling our landfills with mountains of discarded food containers. Hopefully, future delivery container manufacturers will be able to utilize a material that is compostable, biodegradable, or recyclable.

Reducing the cost of a hot meal

When a customer places an order with a delivery company (like UberEATS or DoorDash), a delivery driver must retrieve the meal and transport it to his front door. After computing the delivery fee and the driver's tip, this service may account for 20 to 40 percent of the total meal cost. It is precisely this expense that autonomous vehicles are expected to alleviate. If delivery companies succeed in making home delivery fully autonomous, then they can reduce this cost substantially. By eliminating the human driver from the home delivery equation, autonomous couriers should be able to move all cargo (including hot meals) at decreased prices and with increased efficiency. But a reduction in home delivery fees is not the sole benefit to be accrued once dark kitchens are paired with autonomous couriers. The model may also enable future restauranteurs to avoid many of the operating costs that traditional restaurants have been forced to bear. We'll describe some of these costs now.

Why do restaurants cost so much?

In 2017, the restaurant analytics company Plate IQ examined data for five common west coast food items: hamburgers, burritos, pizza, omelets, and Cobb salads. They found that customers could expect to pay a markup price ranging anywhere from 155% to 636%. Meaning that, though the ingredients in your last cheeseburger only cost the restaurant about $2 bucks, the price tag on the menu read $10 after its 400% markup.

So why is the price so high?

When you buy lunch at your local eatery, you're not just purchasing the food on your plate of course. Your payment is apportioned out to a grand list of operating expenses. Aside from paying rent and maintaining a storeroom full of food and beverages, restaurants must employ a motley crew of cooks, waitresses, cashiers, and busboys. Expenses vary between restaurants and are particularly dependent upon the location of the diner and the sophistication of the menu. But, for most restaurants, the three biggest expenses are:

1. Food
2. Labor
3. Rent & Utilities

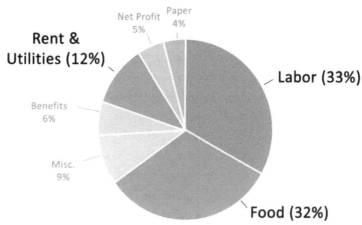

Typical Restaurant Expenses

Figure 45 - Costs vary between restaurants but these stats (compiled by the Family Hospitality Group) approximate monthly expenses.

If we could minimize the cost of these three items then we could reduce the cost of a ready-made meal. Below, we'll discuss how the coupling of dark kitchen centers and autonomous vehicles could make this reduction a reality.

#1: Reducing the Cost of Food

Humans waste a lot of food. In attempting to estimate the severity of the problem, the UN's Food and Agriculture Organization arrived at the staggering estimate that:

One-third of all food produced globally is either lost or wasted.

This waste occurs at every link in the food supply chain—including production, transportation, and consumption. Domestically, things look even worse. In 2012, the Natural Resources Defense Council (NRDC) reported:

Getting food from the farm to our fork eats up 10% of the total U.S. energy budget, uses 50% of U.S. land, and swallows 80% of all freshwater consumed... Yet, 40% of food in the United States...goes uneaten. This...means that Americans are throwing out the equivalent of $165 billion each year...

Analyzing data from the 4,000 US households that participated in the USDA's National Household Food Acquisition and Purchase Survey, Edward Jaenicke of Penn State University concluded:

...the average American household wastes 31.9% of the food it acquires. More than two-thirds of households in our study have food-waste estimates of between 20% and 50%.

Commercial food prep (particularly in non-franchised diners) often results in even more waste. US restaurants throw out approximately 4 to 10 percent of their food before it even reaches the patron's plate. Of the food that does make it to the plate, about 17 percent will remain uneaten.

Such stats are typically not surprising to those who work in the industry. For those who don't, a quick stroll down the aisle of your local diner usually reveals several newly abandoned tables—laden with partially consumed meals and rejected condiment dishes—remnants of the recently completed banquet.

So why do restaurants cook so much food if people don't eat it all?

Perhaps in an effort to manifest the illusion of *value via volume*, restaurant portion sizes have been increasing for decades. In nearly every food category, portions are commonly 2 to 8 times larger than a standard FDA serving size. As the NRDC reported:

Portion sizes have increased significantly over the past 30 years. From 1982 to 2002, the average pizza slice grew 70% in calories, the average chicken Caesar salad doubled in calories, and the average chocolate chip cookie quadrupled.

Many people mistakenly believe that discarded restaurant food somehow makes its way to food banks or homeless shelters. But a 2014 paper by the Food Waste Reduction Alliance insists that only 1.4% of abandoned food is donated to charity. Around 55% of meal leftovers are wrapped up in doggy bags and taken home. But the vast majority of discarded food ends up in a landfill along with the rest of the restaurant's outgoing garbage.

- So how can we fix this mess?
- How can we reduce the amount of waste produced in the transportation, preparation, and consumption of each meal?

Let's consider a few ways that autonomous vehicles might help to alleviate this problem:

- First, the nationwide adoption of an autonomous infrastructure should improve the efficiency of *every* link in *every* supply chain—including those used for commercial agriculture and the foodservice industry. If we can reduce the amount of waste that occurs in the transportation of food from farm-to-fork then we can reduce the cost of food in total.

- Second, future meals will be "made to order"—portioned out as per the preference of the customer. In a fast-food paradigm, there are often three sizes: small, medium, and large. But in restaurants, portion sizes are typically not so easily tailored. Thus, at your local diner, a 300-pound bodybuilder receives the same chicken dinner as a 110-pound yoga instructor—resulting in mismatched portion sizes and more food waste. But, in the coming world of mass-meal production and app-based ordering, meals will be crafted à la carte— according to the customer's requested portion size and dietary parameters. Customers will also have the ability to eschew commonly discarded items—like unwanted side servings, garnishes, and condiments. They'll only be paying for the food that they actually intend to eat.

- Finally, kitchen consolidation should lead to increased meal production efficiency. For example, which restaurant schema do you think will produce more food waste: Five Denny's kitchens—serving five different areas of a city, or one large Denny's kitchen—serving the entire city via autonomous delivery vehicles? A single large kitchen (especially one that is using metrics to forecast demand), should operate more efficiently and produce less waste than several smaller kitchens—operating at independently staffed locales and serving a variable clientele. In a fully autonomous world, the cost to transport a meal *across the street* will be negligibly different than the cost of transporting a meal *across town*. Thus, there will be no benefit in operating more than one chain restaurant in any given region. We'll be talking more about this phenomenon in Chapter 9. But for now, it's likely that the consolidation of multiple kitchens into a single facility will result in increased efficiency and a smaller staff— which should result in yet more savings for the consumer.

Undoubtedly, many more efficiencies will be realized at every step in the food production process. Given the amount of food that America produces every day, even modest reductions in food waste will result in significant gains. The same above-cited NRDC report proclaimed that:

Reducing food losses by just 15% would be enough food to feed more than 25 million Americans every year.

#2: Reducing the Cost of Labor

Labor costs make up the largest slice of the restaurant expenditures pie. Finding people with enough conscientiousness to show up on time and serve *food-with-a-smile* is a constant struggle for restaurateurs. A brigade of workers must be employed, including:

- chefs
- line cooks
- dishwashers
- waitresses
- busboys
- cashiers
- janitors

However, when it comes to dark kitchens, there are no customer tables to wait on. The staff is not burdened with the chore of customer service.

- No bussing of tables
- No swiping of credit cards
- No folding of napkins
- No polishing of silverware
- No refilling of ketchup bottles
- No restocking of dessert display cases

Instead, a dark kitchen's staff need only focus on food prep. The size of their labor force is often a fraction of that of conventional restaurants.

#3: Reducing the Cost of Rent & Utilities

Traditional restaurants often rely on *location* to provide them with a steady stream of foot traffic or access to a regional clientele. However, when it comes to dark kitchens, location doesn't matter so much. Each restaurant is potentially capable of serving every customer who resides in the radius of delivery. So dark kitchen owners don't need to pay exorbitant rental fees to secure prime locales. This realization was best described by the founder and CEO of *Ghost Kitchens Canada*, George Kottas—who operates dozens of dark kitchen facilities. In describing how he starts a new dark kitchen venture, he told CBC Television:

I call up real estate agents and I ask them for the worst location in the city, with the cheapest rent. Because I don't need a storefront [and] I don't open to customers...

Because conventional restaurant selling points (like location and décor) are not a factor for dark kitchens, their facilities can be placed in less desirable or less-trafficked parts of town. And, because they don't feature a dining room, additional savings can be accrued via a reduction in operating costs. In a dark kitchen, there are:

- No plasma television sets
- No sound systems
- No customer-side utilities—like Wi-Fi, dining room heating, air conditioning, and lighting
- No POS (point-of-sale) stations are needed—since all transactions are processed online
- No customer parking
- No store fascia or store signage
- No tables and chairs
- No fancy architectural elements or interior décor

When you take a moment to consider the number of services that dark kitchens do *not* need to provide, then the many potential savings garnered via this business model become salient. Ideally, these savings will be passed on to the customer once autonomous courier vehicles are reliably operating.

The new "fast-food"

When driving down the avenues of any American town, we are typically greeted by the familiar signage of national chain restaurants and fast-food joints. Local boutique restaurants crouch in the periphery of every strip mall. But in an unfamiliar town, they are invisible to us—we can't see their menu from the road. And even if we were to stop our car and sample their food, this sojourn could be costly.

- We'll have to find a parking space.
- We'll have to get out of the car.
- We'll have to sit and wait for them to cook the food.
- We might not like it.
- And, it will undoubtedly be more expensive than the entrées on offer at the fast-food chain across the street.

Given the above constraints, it is typical for us to select the easier, cheaper, and faster option. We too often opt for the default choice and pull our car into the nearest McDonald's drive-through. That's unfortunate because this choice is rarely the healthier option and we're missing out on more diverse dining alternatives.

Thankfully, in the coming autonomous age, drive-through windows will no longer exist. Home meal delivery will be the norm. If you get hungry in transit, you'll simply pull out your cellphone, place your meal order, and then request for it be delivered to your destination address. Or you might rendezvous with your incoming meal at a pick-up location. Most importantly, your future restaurant choices will not be limited to the half-dozen fast-food chains that occupy the right side of whatever street you're driving down. Instead, *every* restaurant in your delivery radius will be happy to cater to your needs. Since every meal order will be placed via app, every restaurant will show up on your cellphone as a potential dining option. This should equalize the playing field—allowing smaller eateries to compete with the larger chains. Restaurant competition is currently heavily dependent upon restaurant location—once a chain restaurant secures a position in a neighborhood, they leverage their existing brand recognition to lure local traffic. The power of these brands is evidenced by the staggering number of franchise locations that exist worldwide. Below, the top ten chain restaurants are listed beside their 2019 location count.

1. Subway - 42,431 locations
2. McDonald's - 37,855 locations
3. KFC - 20,404 locations
4. Burger King - 16,859 locations
5. Pizza Hut - 16,796 locations
6. Domino's - 15,000 locations
7. Hunt Brothers Pizza - 7,300 locations
8. Taco Bell - 7,000 locations
9. Wendy's - 6,490 locations
10. Hardee's - 5,812 locations

In the online meal delivery model, every restaurant will not only be competing for the traffic that is in proximity to the dining room, they will *also* be competing with every restaurant in the customer's autonomous delivery radius. More competition means lower prices for consumers, but it also means more meal options. Hopefully, some of these options will be of a healthier variety.

There is a natural tendency to assume that, just because food is "fast" this implies that it is not healthy. Given the nature of the current US fast-food industry, this assumption is mostly correct. But as autonomous delivery systems come online, we may finally have a way to bring the convenience of "fast-food" to *every* meal category—not just burgers and burritos.

Ideally, this technology will allow us to redefine what we mean by "fast-food." In addition to costing a fraction of what meals cost today, the food produced via this process has the potential to be healthier and more nutritious than any food product ever manufactured in large quantities.

"Is the Kitchen Dead?"

It happened in January of 2015. This was the month when retail sales for restaurants and bars overtook grocery store sales. Meaning that, for the first time since the US Census Bureau has been tracking consumer food spending, Americans spent more money "eating out" than they did "eating in."

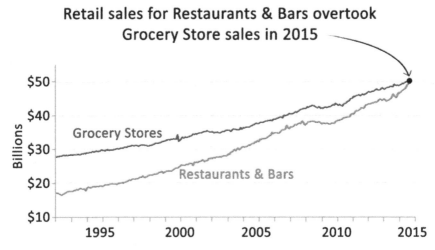

Figure 46 - Restaurant and bar sales overtook grocery store sales for the first time in 2015. (Source: Data from the US Census Bureau's 'Business and Industry' monthly report from January 2015.)

In 2018, the Swiss investment bank UBS published an influential report titled "Is the Kitchen Dead?" Here, UBS estimated that the global online food delivery market would grow tenfold over the next decade—from $35 billion to $365 billion by 2030. They write:

…by 2030 most meals currently cooked at home [will instead be] ordered online and delivered from either restaurants or central kitchens… [A professionally] cooked and delivered meal could approach the cost of home-cooked food, or [even] beat it [if the customer's] time is factored in… Powering this are lower meal production costs, increased logistics scale, and strong demographic trends. Shared "dark kitchens" cut meal preparation costs, and robot chefs industrialize the

process further. At scale, unit costs of delivery drop by 50%, and even more via drones and droids.

Currently, the total cost of meal production is inflated by the cost of transporting food from the farm to the food hub, and then to the kitchen of your favorite local restaurant. If you order dinner via a meal-delivery service, their driver must transfer it a fourth time—from the restaurant to your porch. In considering the many hands that are required to make farm-to-fork delivery possible, it's easy to understand why restaurant margins are so narrow and why restaurant meals are so expensive.

Autonomous vehicles will allow us to chip away at food transportation costs—at each link in the supply chain. As prepared meal manufacturers and dark kitchens continue to expand and utilize autonomous technology, the price of a home delivered meal will continue to drop. Just as it is currently uneconomical to grow your own vegetable garden, it may someday be uneconomical to cook your own meals. The art of "home cooking" could become just another anachronistic curiosity—a quaint hobby not unlike archery or pottery.

We will always have hobbyists or artisans among us—people who engage in the culinary arts for the pure joy of it. But most people of the future won't be motivated to *learn how to cook* any more than most of us currently feel motivated to *learn how to sew*. The chore will become almost entirely divorced from the daily affairs of future man. If we succeed in completely outsourcing meal production, our only involvement with food will be to put it in our mouths and chew. *Kitchen work* will become a foreign concept to future Americans—just as *farm work* is to most of us now.

For many, this news is cause for celebration. As stated before, the average American devotes five years of life to the chore of attaining and eating food. If this burden can be alleviated on a nationwide level, then *centuries* of time can be freed up—hopefully, to be filled with more productive pursuits.

In an interview with the New York Times, NPD Group trend researcher Harry Balzer summed up the coming revolution this way:

A hundred years ago, chicken for dinner meant going out and catching, killing, plucking and gutting a chicken. Do you know anybody who still does that? It would be considered crazy! Well, that's exactly how cooking will seem to your grandchildren: something people used to do when they had no other choice.

We don't know exactly how the home delivery of cooked meals will be automated in the future. But it is clear that many of us have abandoned the chore of "home cooking" already. However, even after all of our home kitchens have been converted into kitchenettes, and all of our fast-food franchises have retreated into our iPhones, the ritual of *communal eating* will remain a persistent facet of the human experience. To satisfy this most ancient of social compulsions, we will always need public restaurants. However, the *restaurant experience* itself will change.

Ch. 6: How AVs will Change the Restaurant Experience

In consideration of the meal-delivery technologies discussed in the previous chapter, one might conclude that the *hometown diner* is going away. It's not. But clearly, the restaurant industry will be disrupted when autonomous vehicles are able to deliver meals in minutes. The introduction of such technology will surely change the way food is consumed. In this chapter, we'll ponder the following questions:

- Given that autonomous vehicles will be able to deliver *any meal*, at *any time*, and to practically *anywhere*, then what effect will this have on local cafes, bars, and diners?

- Once food-preparation and delivery are both automated and the commoditization of "restaurant food" is complete, then what type of service will local restaurants be providing?

- What will restaurants be like in the future?

The Restaurant of the Future

The more gregarious among us will be relieved to hear that the birth of autonomous vehicles won't mark the end of restaurants. There will always be a need for public venues—in which people can socialize while consuming food and beverages. Currently, restaurants cater to this desire by providing tables, chairs, and cooked-to-order meals. But with the advent of mobile food-ordering apps, patrons now have access to the menus of every restaurant in the city—not just the menu of the restaurant they happen to be sitting in. And given the speed of the emerging autonomous meal-delivery infrastructure, it will someday take about the *same* amount of time to order a chicken dinner from a restaurant on the other side of town, as it does to order one from the waitress standing in front of you.

Currently, restaurants don't allow outside food to be consumed at their tables. But future restaurant business models may not be so adamant about this policy. Some will welcome such 3rd party deliveries and may alter their facilities to accommodate the transfer of meals from courier to patron.

Let's consider three types of such venues now.

Venue 1: Open-air Eatery

The proceeding rendering features an open-air eatery and bar. Along the facility's public-facing side are located several numbered dropzones—at which patrons can receive incoming food deliveries from autonomous couriers. Recall that, in the future, all of such dropzones will be assigned a unique address—their coordinates will be registered in a national database which courier vehicles will use to find the location at which their cargo is to be dispensed. Given the ease by which future meals will find their way to hungry customers, venues like this might not even bother with the expense and complexities of onsite food preparation. Instead, they might monetize the establishment by other means—like via drink sales or perhaps by charging an entrance fee at the door.

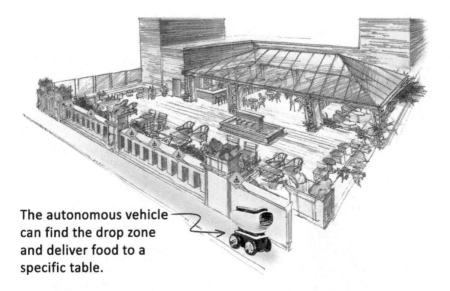

The autonomous vehicle can find the drop zone and deliver food to a specific table.

Figure 47 - This open-air restaurant features numbered dropzones—at which autonomous vehicles can deliver meals.

Venue 2: Outdoor Recreation Area

Once autonomous delivery is perfected, the phrase "eating out" need not refer to patronizing a brick-and-mortar facility. It will be possible to receive meal deliveries at just about any commonly trafficked location—including public parks, beaches, or campsites. In the following rendering, dropzones have been established near picnic tables by a beach. To retrieve his lunch order, the requesting customer need only rendezvous with the incoming autonomous vehicle at the agreed upon dropzone location.

Figure 48 - A customer uses a meal-delivery app on his cellphone to rendezvous with an autonomous meal courier at a dropzone in a park.

Venue 3: Conventional Restaurant with meal-delivery option

Aside from open-air eateries or outdoor recreation areas, conventional restaurants (with walls, tables, chairs, and bathrooms) will probably always comprise the majority of dining venues. However, these locations might not necessarily staff a full-service kitchen. Instead, patrons might be encouraged to order meals via delivery app and use the establishment's address as the drop-off point. Let's describe how this process might work:

STEP 1 **STEP 2** **STEP 3**

- **Step 1**: Patrons would first order their meal via the typical app-based channels. An autonomous courier would then transport the meal to the facility's dropzone—perhaps depositing it into a portal near the restaurant's kitchen entrance.
- **Step 2**: Then, a "food presenter" (not a cook) could receive the incoming delivery. He'll extract the meal from its container and arrange the contents onto a plate—perhaps adding condiments, appetizers, or side dishes.
- **Step 3**: Finally, a waitress would bring the meal to the customer's table, along with silverware, napkins, and beverages.

In such a model, most of the traditional "dining experience" is salvaged. From the customer's perspective, not much has changed—he placed a meal order and the waitress brought him a plate of food. Yet, a kitchen brigade is not required in this model. Only a skeleton crew would be needed to arrange meals, bus tables, and serve drinks.

Someday, when the autonomous infrastructure is complete, such dining schemas could become the norm. The majority of brick-and-mortar eateries may elect to outsource the burden of food-prep to large dark kitchen centers. They might even nix their menu entirely—instead choosing to monetize the facility by some other means; perhaps differentiating themselves (not by food type), buy via experiential traits like location, interior design, clientele, rides, games, live music, or theme.

The Rise of Theme Restaurants and Experiential Dining

So-called "theme restaurants" — eateries promising a unique *dining experience* via trendy décor — have been around for decades. A few familiar national chains include:

- Medieval Times – A restaurant that also features jousting stuntmen.
- Chuck E. Cheese's – A restaurant that also features rides and games for children.
- Dave and Buster's – A restaurant that also features rides and games for adults.
- Hooters – A restaurant that also features scantily clad waitresses.
- The Hard Rock Café – A restaurant that also features rock 'n roll memorabilia.
- The Rainforest Café – A restaurant that also features animatronic animals.

It is the value-add (following the phrase *"also features"*) that may help to differentiate future brick-and-mortar locations from their online competitors. If the efficiencies of the autonomous age ever succeed in rendering the chore of onsite cooking superfluous then future restaurateurs may elect to focus their efforts into the development of a more elaborate *restaurant experience*.

Not all eateries of the future need be as gaudy as our above-cited examples. Many restaurants will offer a more staid or serene experience—like a table near a cove, a garden, or a babbling brook. Some will feature more conventional entertainment—like acoustic guitar music or a lively local bar. Still, other restaurants may not offer much more than any contemporary diner. They'll have clean tables, clean bathrooms, beverages, ketchup bottles, and may primarily function as a local gathering spot—a place for people to meet, eat, and converse outside of the home.

Future facilities may not cook much onsite—instead encouraging customers to order meals via autonomous courier. But this doesn't nullify their raison d'etre. In practice, most restaurants have never *only* existed to sell food anyway. They have always provided a dual-use option. Friendships are forged, business deals are made, marriages are proposed, and books are written atop restaurant tables. Even in a future world (where a robot can sprint a turkey sandwich to your table in five minutes flat), most people will prefer to spend at least *some* of their time in communal dining environments. Restaurants function as social hubs that cater to this need to commune. Such transmissions are only possible when we sit in proximity to a circle of friends—preferably in a facility that fosters an inviting ambiance. Such a service can't be packaged into a delivery box. And it's *exactly* this type of service that brick-and-mortar restaurants are in a position to provide. As Nico Larco of the University of Oregon's Sustainable Cities Initiative put it:

In the past, I might have gone to that strip mall down the street because it was the closest place for me to buy something. But that might change in the AV world. What might become more important is the quality, the vitality of a place, the buzz, the synergy. If I need some commodity, I can buy things from Amazon. [If] I'm going out, [then I'm going out] for the experience. Architecture and urban design used to be the containers of commerce. [In the future,] they might very well become the generators of it.

When Robots Aren't Enough

No matter how fast autonomous meal-delivery vehicles get, they will never succeed in replicating every service that regional eateries are capable of providing. If you're walking down the street, it will always be quicker and easier to buy a hot dog from your local street vendor than it will be to ask a robot to deliver it.

Figure 49 - Even after autonomous delivery bots are ubiquitous, some local meal providers will always be in a better position to provide immediate service to customers prone to spontaneous buying decisions. (Photo: Hu Totya)

In vivacious venues with high foot traffic (like sporting events, business parks, and shopping districts), individual vendor sales will remain the norm. Businesses reliant upon location and spontaneous buying decisions will forever have a market niche—particularly when it comes to chilled single-servings like ice cream and alcoholic beverages. For such delicacies, it's difficult to see how the use of autonomous couriers would increase the efficiency of the transaction. Cocktails will probably always be mixed near the place where they are drunk. However, some bars may adopt other forms of automation.

The Italian robotics company MakrShakr claims to have developed the "first robotic bar system"—able to "create limitless cocktail recipes…assembled and served by robotic arms." Anyone who has ever tried to order a drink in a swanky big-city nightclub might welcome the addition of a couple "robotic arms" to those of the insufferable bar staff. But for more provincial establishments, the new mechanical hires probably wouldn't be favorably received by the local barflies.

Figure 50 - In the top image, MakrShakr's mechanical bartender serves drinks. In the bottom image, a female bartender prepares limes. (Photo by Dave Dugdale.)

In glancing at the preceding image, it is clear that the lower panel features an employee who is capable of providing an ancillary service—one that goes beyond merely pouring alcohol into a cup. Though a robot bartender may make more drinks-per-second than its female counterpart, it won't be able to laugh at your jokes. Thus, even after mechanical bartenders are available on the cheap, we may not see an immediate end to the human-powered models.

Yes, even after autonomy has been mastered, many establishments will elect to remain just as they are now. They'll mix their own drinks at their own bar, and they'll prepare their own dishes in their own kitchen. Though such meals might come at a higher cost, they will be crafted via the skilled hand of an artisan cook, or poured by a cocktail waitress endowed with feminine charms.

Classical fine dining will always have an appeal. Some people enjoy the experience of sitting down at a banquet table (laden with polished silver and cloth napkins), and being waited on by an attentive human—especially one temporarily cast into the role of *indentured servant*—at least for the duration of the meal.

The *fine dining experience* is just another type of *experiential dining* of course. It allows the patron to be "king for a day"—to bask in a transient milieu of luxury. This fusion of sublime presentation and social decorum is not so easily commoditized or automated; some meals contain ingredients that cannot be strapped to the back of a robot. Given that the constituents of the dinner are not the primary impetus of such positive emotions, there will always be those who are willing to pay for the *experience* of fine dining.

Ch. 7: How AVs will fix the Housing Crisis

About those Google Engineers who lived in their RVs...

Starting around 2014, several news outlets began running stories about tech workers who had failed to find affordable accommodation around Mountain View, California—the home of Google's corporate headquarters. To avoid the abhorrently high cost of rent, some employees elected to reside atop the company's parking lot in RVs and camper vans.

Figure 51 - Ex-Googler Matthew Weaver reportedly lived in the parking lot of Google's Mountain View campus from July 2005 to August 2006.

It's not clear how many Google employees homesteaded on the black asphalt plain, nor how long the racket lasted. But California zoning restrictions do not permit RVs to camp on corporate lots. So, assumedly, Google eventually managed to convince the squatters to move along. However, as many Silicon Valley residents will attest to, it is still easy to find local tech employees sleeping in their cars on the streets of Mountain View. At the time of publication, you can drive down Crisanto Avenue (just two miles south of the Google campus) and find an array of RVs resting quietly under the shady redwood trees.

In contemplating the plight of these well-employed (but homeless) men, several questions come to mind:

- Why are these highly-skilled people living on the streets?
- Why are the employees of the most financially successful company on the planet sleeping in the company parking lot?
- Is it just to save money on rent?
- Why is the rent so damn high?

"The rent is too damn high!"

In 2019, rental prices in Silicon Valley hovered around $3,300 for a pad with an average size of around 828 square feet. For many Googlers (some just out of college), paying nearly $40,000 a year to rent a single room in a tiny apartment was difficult to rationalize. So it's easy to see why many of them succumbed to bohemianism.

Housing costs comprise the largest financial burden for the average American—accounting for over 32.8% of monthly expenditures. The cheeky reports about the Google engineers living in their cars made for entertaining headlines. But the stories of these men are emblematic of a very serious housing shortage. In west coast boomtowns (such as Silicon Valley, neighboring San Francisco, Los Angeles, San Diego, and a dozen other cities along the US coastline), prohibitively high rental fees have caused many to flee the state. Home prices have been going up across the country (especially in New York, Boston, Seattle, and Miami). But in California, the real estate rollercoaster has been particularly dizzying.

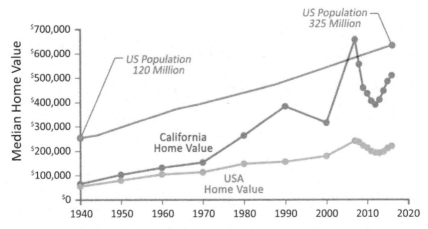

Figure 52 - Median home prices have been on the rise across the US, but are particularly precipitous in California. (Source: U.S. Census Bureau (2000) Historical Census of Housing Tables Home Values. Prices shown in 2019 dollars.)

We are often quick to blame the increase in housing costs on the increase in our nation's population size. The US population has indeed grown. In fact, in the last six decades, it has doubled—moving from 165 million citizens in 1955 to 325 million in 2020. Multiplying the size of a nation's population surely increases housing demand. Thus, we might infer that the rent is so high because we're running out of room to build new houses. But this isn't the case. America has a lot of land.

Of the 1.9 billion acres that comprise the 48 contiguous states, over 80% of the territory is merely devoted to three uses:

1. Pasture and Rangeland
2. Cropland
3. Forested regions

The business of most American land is *agriculture*. When pastures and feed-producing cropland are added together, an incredible 41% of this sum is solely devoted to rearing livestock—providing the nation's 100 million cows with spacious grazing opportunities. Untouched forest comprises the second most sizeable chunk of US land—consisting of federal and state parks, as well as corporate and privately owned timberland.

169

The following diagram should help you to appreciate the immense scale of these three interests. And it may become easier to understand why 47% of US census blocks have not a single soul living on them.

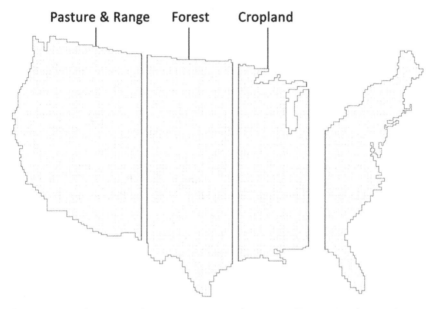

Pasture & Range   Forest   Cropland

Figure 53 - If we squash every acre together according to land-use, then it becomes strikingly evident that the majority of US land is merely devoted to agriculture or untouched forest. (Source: Diagram based on calculations by Dave Merrill and Lauren Leatherby for Bloomberg Businessweek.)

Clearly, America has a lot of land lying around. So how come nobody is living on it?

People tend to cluster into tight collectives. According to a 2015 report by the US Census Bureau, 62.7% of Americans live in cities—which occupy only 3.5% of the total US land area.

· One dot = 7,500 people

Figure 54 - Most of the U.S. population is tightly clustered around urban centers. In this map, each black dot represents 7,500 people. (Source: Original rendering by the U.S. Census Bureau based on 2013 Population Estimates and 2010 Census data.)

As the Census Bureau so aptly put it:

The population density in cities is more than 46 times higher than the territory outside of cities. The average population density for cities is 1,593.5 people per square mile, while the density outside of this area is only 34.6 people per square mile. Population density generally increases with city population size.

As a city grows, so too does its population density. Rent is high, and houses are expensive because people are forced to compete for residences that have proximate access to jobs and services. The path we tread between our home and our office is typically the most familiar route in our lives. Urbanites must dutifully traverse this circuit five days a week. I's a costly trek—in terms of both time and fare. Homeowners attempt to minimize these costs by purchasing residences that are closer to their place of employment. Because this proximal land (by definition) is more scarce, it is also more expensive. As a city center continues to thrive, consumers must compete for a dwindling supply of accessible dwellings—thus driving up the cost of housing.

The housing crisis is a persistent ailment, against which cities have attempted many remedies. Often they'll try to increase land accessibility by improving public transportation, widening roads, or adding parking lots. When things get really tight, they'll try to increase the housing supply by building vertically or by decreasing apartment minimum square footage requirements. A 50-square-foot Manhattan 'studio' apartment on the Upper West Side could rent for over $500 per month in 2019.

Such solutions usually don't remedy the high cost of housing. Nor do they ever really reduce commute times—at least not for very long. In bustling cities, new homes still cost too much, apartment rent is too high, and commute times are worse than they ever have been.

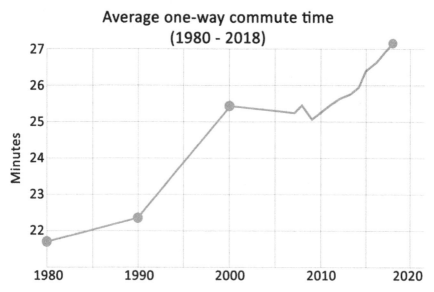

Figure 55 - The average US commute has reached a record high of over 27 minutes one way. (Source: US Census Bureau Estimates from 1980 to 2000 are from the decennial Census. Estimates from 2007 to 2018 are from the American Community Survey. Data originally compiled by The Washington Post.)

Thus far, city dwellers have tolerated the encumbrances of increased urban density in exchange for prosperity and proximity to jobs and services. But, perhaps soon, self-driving technology may allow us to have growth without gridlock.

In this chapter, we'll discuss three ways that autonomous vehicles will free us from this imposed compromise by:

1. making land more accessible,
2. freeing up land for development,
3. and, relieving urban congestion via gains in transportation efficiencies.

Part 1. How AVs will make land more accessible

When it comes to calculating home value, real estate agents are quick to cite their three-word mantra: "location, location, location." Meaning that the *price* of a home is primarily determined by the *location* of the home and that this consideration warrants more scrutiny than any other variable; far surpassing other features like square footage and lot size. However, in the coming age of autonomous vehicles, the efficacy of this rule will be diminished. As Urban Planner Paige Pitcher said at the 2017 Tedx conference in Ogden, Utah:

...that rule is flawed. Real Estate is not about location. It's about transportation. It's about 'how you get there.' And that's [why even] small changes in the automobile industry can mean huge things for real estate. And, let me be clear, the autonomous car is no small change.

To understand how much change autonomous cars will introduce to the world of real estate, let's consider a typical home buyer's journey. Imagine that you're searching for a new house or apartment. Often, when initiating this process, people will first identify the location of their current place of employment, and then they'll consider the number of miles that they're willing to tolerate during their morning commute. This resultant distance becomes the radius in which their new home must reside. Because the journey between home and office is so frequently trekked, it is in our interest to minimize this variable, while still attempting to find our "dream home" within its bounds. This is why "location" has traditionally been such a crucial factor in the home-buying calculus. But now, let's tweak a few parameters. Ask yourself how the following modifications would affect your home-buying process:

- What if you didn't have to drive to work in the morning? Suppose that you owned a self-driving car and it spun your steering wheel for you—while you relaxed in your seat.
- What if you didn't have any "soccer mom" duties? E.g. suppose that autonomous taxis were capable of executing all of your familial transportation obligations—like dropping the kids off at school, or at church, or at Little League practice.
- Further, suppose that the autonomous home delivery infrastructure has been completed. Couriers casually drop off and pick up parcels from your home via a mechanized process—thus eliminating the need to ever run errands or go grocery shopping again.
- Finally, suppose that (on some stretches of road) your self-driving car could safely reach speeds of nearly 100 miles-per-hour—thus making long-distance commutes feasible.

If the above-described conveniences existed today, how would they affect your home-buying decision? How would improvements in transportation technology expand the radius in which you are willing to buy a home or rent an apartment?

Currently, the average American lives about 16 miles from work. But soon, the travel distances and durations that people can tolerate will be extended by the many efficiencies achieved via the coming autonomous infrastructure. Remember, in this future, there will be no rush hour traffic, no parking lot labyrinths, no fender benders or looky-loos, and someday your self-driving car won't even need to stop at intersections. Such innovations will allow residents to extend the expanse between their homes and the city center.

The Standard Urban Model dictates that the cost of land *decreases* as the distance from the city center *increases*. Not every city on Earth precisely conforms to this curve. But, as a general rule, housing costs drop as we get less *urban* and more *rural*.

Figure 56 - The Standard Urban Model dictates that land prices fall as the distance from the city center increases. (Source: Plot by Alain Bertaud of the NYU Stern Urbanization Project.)

Historically, people have been forced to tolerate higher housing costs for low commute times. Or they have endured longer commute times for reduced housing costs. But, given that autonomous vehicles will be so apt to carry us the extra mile (for a negligible increase in fare), future homebuyers will not be so strictly constrained by mere distance. Self-driving technology will allow them to purchase homes that are much farther from work, school, or downtown fun. Builders will be able to construct homes on the outskirts of the city—on land that is currently *too far* from the city center to be practical for daily commuters. In theory, this newfound accessibility to peripheral residences should increase the supply of homes—thus lowering the cost of housing.

The number of minutes that future urbanites will add to their current commute times will depend upon the degree of efficiency that future autonomous vehicles manage to attain. The cities that succeed in constructing an autonomous infrastructure will likely see a rapid increase in the number of "super commuters" — people whose daily commute takes over 90 minutes one-way. The number of drivers who match these criteria has been steadily growing since 2009—recently increasing at a rate of about 3.7%

175

annually. But when self-driving cars start appearing on our freeways, this number is expected to multiply. As Daniel Piatkowski (Professor of Community and Regional Planning at the University of Nebraska) said:

> **[We're going to see the rise] of the super commuter... [People who travel] into the city every morning—[traversing distances that] would have been hours away... But in a driverless car [the trip would only take] an hour at most.**

Once people can comfortably commute long distances via autonomous vehicles, then the horizons of their hometowns will broaden. No longer will they be forced to compete for housing merely because of its proximate distance to jobs and services. Instead, they'll be able to settle in regions of the country with lower urban density and (most importantly) with lower housing costs.

Part 2. How AVs will dump more land on the market

Aside from increasing the supply of accessible homes on the city outskirts, autonomous vehicles will also increase the supply of homes on offer downtown. This becomes evident when you consider how much of our current urban infrastructure exists solely to accommodate conventional automobiles.

In this section, we will consider three areas from which urban land will someday be reclaimed:

I.   From parking lots
II. From garages
III. From other soon-to-be-defunct businesses

I. Reclaiming land from parking lots

In some urban zones, parking lots may account for over one-quarter of land use, and there may be 3.5 parking spaces for every one car. To better visualize the amount of space we're talking about, consider the following image:

Figure 57 - In this California commercial retail district, the parking lots have been colored in black.

The majority of the developed land in this commercial retail district consists of parking lots and multi-story garages. Such layouts are common in many US urban centers—particularly those that evolved after the conception of the automobile. If society succeeds in adopting autonomous ridesharing vehicles (robo-taxis) as the primary means of transportation then citizens will no longer be obligated to park their cars at every destination they visit. Robo-taxis don't park. After they drop a passenger off at his destination, the taxis simply return to the road to find a new customer. Retail parking lots, as we know them now, will no longer be required. This realization prompts the question:

"What are we going to do with all of these parking lots?"

In cities that welcome the autonomous renaissance, developers will be able to rethink the way that commercial buildings and apartments are designed.

No longer will architects be forced to surround their complexes with black asphalt pastures. Instead, our buildings (office parks, stadiums, hotels, apartments, shopping centers, etc.) won't need any parking facilities at all. As Nico Larco (professor of architecture and environment at the University of Oregon) told Newsweek in 2018:

…there are more than a billion parking spaces in America… If [autonomous vehicles] are doing all the driving, we can get rid of 90% of them.

Nobody knows exactly how many parking spaces exist on US soil. There are at least 500 million off-street parking spaces. But if you consider curbside parking, garages, and parking lots, then the number of spaces may be well over a billion. Regardless, freeing up *any* percentage of parking lots will dump *a lot* of land onto the US real estate market. Most importantly, many of these parcels will be located near the city center. Meaning that builders will have access to millions of acres of developable graded land—all located in prime real estate locations.

In a 2018 report by the *Research Institute for Housing America*, data scientist Eric Scharnhorst attempted to put a dollar amount on the value of parking space land in five cities: New York, Philadelphia, Seattle, Des Moines, and Jackson. His team estimated that the lots were worth around $81 billion. If this figure is reflective of the value of lot space in only five US cities, then the national figure is undoubtedly astronomical. In referencing his data, Scharnhorst wrote:

It's no secret in the development world that parking lots are *land banks*—just waiting to be turned into something else.

Eventually, some of this land will be reclaimed and turned into new residential units. Increasing the supply of units should cause real estate prices to drop. In theory, we should see a reduction in the cost of housing— wherever autonomous vehicles are welcomed.

II. Reclaiming land from residential garages

In 2017, the US Department of Transportation estimated that there are about 1.88 vehicles per US household. This is why even the humblest homes of suburbia must be built to accommodate at least two (sometimes three or four) cars. Often these vehicles spill out of the garage and find a new home on the driveway. As more tenants move in, the cars may burst forth yet again—flowing across the sidewalk and mooring along the curb.

Unfortunately, the avenues of our neighborhoods are too often devoted to the harboring of marooned automobiles. The proud facades of our homes are perpetually obstructed by this curbside flotsam and jetsam. The strategy to deal with this problem thus far has been to affix ever broader garages to each residence. This approach has resulted in some unfortunate architectural abominations.

Figure 58 - Because car ownership is so prevalent in the US, homes are often designed to accommodate several cars per family—resulting in overbearing garages and driveways.

Thankfully, when robo-taxi usage becomes commonplace, new homes will no longer need to be equipped with such unsightly stables. Robo-taxis are never garaged at their rider's residence of course. So new homes will no longer be built with garages. Removing this construction requirement will result in immediate savings. For example, the house in the following diagram features a typical 24' two-car garage—which consumes 576 square feet of lot space. When the driveway is included, this sum is often doubled.

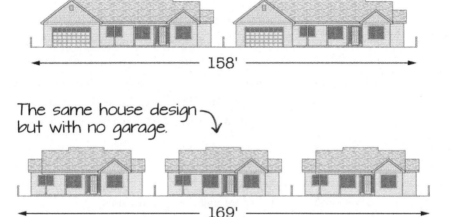

The same house design but with no garage.

Figure 59 - Both tract home designs in this image have the same amount of livable space. But when garages are not required, nearly three homes can fit on lots that currently only hold two.

Devoting this much square footage solely to the chore of sheltering vehicles is a costly and wasteful use of developed land. But notice, in the second image, we have drawn the same tract home sans garage. Here, we are able to fit nearly *three homes* on a lot size that currently only accommodates two. Removing the garage from the floorplan will allow developers to build more homes per square mile *and* reduce the cost of construction—all while maintaining the *same* amount of livable space.

III. Reclaiming land from defunct businesses

Next time you're driving down your local commercial strip, take a moment to consider the goods and services that the businesses have on offer. Then, for each store, ask yourself if autonomous vehicles are likely or unlikely to render its business model obsolete. What you might discover is that most of the stores currently operating in your hometown will probably not thrive in an age of robo-taxis and hyper-efficient home delivery.

- Consider *automotive services*. In 2020, the National Association of Convenience Stores estimated that there were 121,998 gas stations in America. In well-trafficked urban centers, it's not uncommon to find several gas stations per mile. But in the future, self-driving cars will not refuel for gas at individual stations (if they use gas at all).

Future ridesharing vehicles will fill up their tanks (or charge their batteries) in strategically located fueling facilities—dedicated hubs that will service thousands of autonomous vehicles per day. Thus, the land that you currently see beneath every gas station in your town will one day be vacated. A similar fate awaits many other auto services like car washes, car dealerships, auto repair shops, and auto supply stores. The land under each of these shops will be up for sale as well.

- Let's try another example—*restaurants*. In Chapter 5, we discussed how autonomous vehicles will transform meal delivery. Large kitchen facilities will prepare millions of individual meals each day and autonomous couriers will service every home in the given region. If you count full and limited-service restaurants and cafeterias, then there are about 400,000 brick-and-mortar eateries in America, and around 40,000 grocery stores and supermarkets. It's not clear how many existing food retailers will survive the transition to an autonomous economy. But when food can be summoned in minutes via app, we assume that their numbers will decline as well. Thus, much of the land that these restaurants and supermarkets currently sit on may be vacated.

- When it comes to *retail* shops, the National Retail Federation estimates that there are over one million establishments across the US. Of these, 152,720 are convenience stores (80% of which also sell gas). Of the top 50 US online retailers, nearly all of them operate brick-and-mortar stores. Currently, online transactions make up just ten percent of all retail sales. But recall in Chapter 4 we described how homes of the future will come equipped with a dropbox in which autonomous couriers will rapidly deposit all newly purchased items—possibly in under thirty minutes. So if a customer can get a box of cereal delivered to his home with just a few taps on an iPhone, then why would he walk to the convenience store to get it? Why would he go to the liquor store, or the cell phone store, or the bookstore, or the mall? It seems likely that many of these businesses will not survive the conversion to an autonomous economy. The land on which they reside may be open to new development in the future.

We'll talk more about such changes in the coming chapters. For now, it's clear that the conventional retail model will someday be threatened by the introduction of autonomous delivery technology. However, in considering the coming apocalypse, IHL retail analyst Greg Buzek retorted:

> **Retail is still very much about location, location, location to consumers. [In the future, existing] stores may take on more of an ecommerce fulfillment role and radically change formats from today, but their advantage is their proximity to the consumers they serve.**

Indeed, "proximity to the consumers" will be an advantage for a while longer. (Maybe a long while longer.) But when autonomous couriers are finally capable of delivering all retail orders in minutes, then existing retail locations cannot simply convert their showrooms into fulfillment hubs. Storing products in smaller regional processing centers would not significantly reduce delivery times once the autonomous infrastructure is perfected. Instead, it is perhaps more probable that most local stores will eventually suffer the same gradual obsolesce as the local bowling alley or the local drive-in movie theater.

Figure 60 - In the 1970s, there were over 2,400 drive-in movie theaters in America (about 25% of all US movie screens). By 2018, only 320 drive-ins remained in operation.

When confronting technological disruption, economists have typically been quick to exclaim, "New jobs will come! New jobs will come!" Such sentiments have not only assuaged the fears of those who worried that the end is nigh, but they have also been historically correct. Perhaps this track record will hold. After all:

- Can't the drive-in movie theater just be converted into a strip mall?
- Can't the strip mall just be converted into a big-box store?
- Can't the big-box store just be converted into a spaceport?

Perhaps. But it seems unlikely that conventional buying behaviors will be indefinitely recapitulated with each emerging marketplace instance. Given the ease by which goods and services will be whisked to our homes in the future, it is perhaps more likely that tomorrow's consumers won't require direct access to regional storefronts. Instead, they'll just shop online and get their purchase delivered.

As the efficiency of home delivery continues to improve, the necessity for one to move one's body to a brick-and-mortar store will diminish. If this is to be the progression of retail sales, then it's difficult to see how regional businesses will stay afloat in the future autonomous economy. The buildings and strip malls that currently house these stores will one day be vacated. There may come a day when the streets of America's once thriving urban districts look more like the remnants of a strategic bombing campaign than a shopping center.

Figure 61 - We don't know how many brick-and-mortar retail chains will still exist after rapid autonomous transport and home delivery becomes an option for consumers.

The acres on which these stores currently reside will be dumped onto the real estate market—further increasing the supply of available land. Perhaps future economists will come to refer to this period as "The Great 21st Century Land Dump."

Part 3. How AVs will relieve urban congestion

In the previous two sections, we have described how autonomous vehicles will simultaneously make land more accessible *and* potentially introduce millions of new acres to the real estate market. Basic economics dictates that increasing housing supply will decrease housing costs. But this conjecture may be too simplistic to describe the behavior of humans in cities. It fails to account for other metrics like crime, racial demographics, increases in the standard of living, and the many logistical difficulties that come with increased urban density.

Merely increasing the supply of housing units does not always result in a reduction in housing costs. And adding *more* livable units to a city does not always increase the number of people who want to live there. Many are surprised to learn that Manhattan and Brooklyn actually hit their population peaks before 1940.

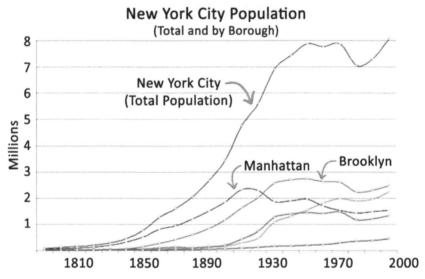

Figure 62 - While the total 2019 population of New York City hit an all-time high of 8.6 million, the boroughs of Manhattan and Brooklyn peaked in 1910 and 1940. (Source: Originally rendered by Julius Schorzman using data from nyc.gov.)

Like bacteria hitting the wall of a petri dish, city populations cannot maintain geometric growth forever. There is an upper bound to the amount of urban congestion that people are willing to tolerate. The act of merely *adding more housing units* may not necessarily lower housing costs in already congested urban areas. Some cities just reach a point where they *feel* crowded—where the tedium of getting from Point A to Point B becomes tiresome and traffic becomes too much to bear. But what is it *exactly* that incites this sensation of *crowdedness?*

In large part, the degree of congestion experienced in a city is a function of the efficiency (or lack thereof) by which people travel through the city's corridors. When these corridors become overcrowded, residents experience discomfort which eventually prompts them to flee. Thus, if we are to bring housing prices down, not only must we increase the number of accessible housing units, but we must also improve the efficiency of the transportation conduits that lie between these units.

In this section, we'll describe four ways by which autonomous vehicles will relieve urban congestion and hasten the flow of people and parcels down the streets of the future.

I. Autonomous vehicles will use the road more efficiently than conventional vehicles

With the staid hands of A.I. at the wheel, the cars of the future will be better at managing the frenzied kinetics of morning rush hour. Recall that robo-taxis don't need to be parked, they don't get lost, they don't get road rage, they don't stop for gas, and they don't get into traffic accidents. But not only will autonomous vehicles make the morning commute faster and safer they'll also use the roads more efficiently. Below, we'll list three reasons why we believe this to be the case:

- First, as their name implies, robo-taxis allow multiple people to share a single ride. In Chapter 3, we discussed several possible seating configurations for future ridesharing platforms. Some of these vehicles will function much like commuter busses while others will be designed to offer door-to-door service to three or four people. Seating arrangements and occupancy levels will vary depending on the time of day and the demands of the given area. But undoubtedly, such vehicles will be able to transport more riders per car than the capacity currently achieved via a *personal car ownership* paradigm. Instead of four office workers driving four conventional vehicles to work, there will be only one vehicle—picking each of them up and dropping each of them off via an algorithmically optimized route. This will decrease the total number of cars on the road at any given time—thus reducing road congestion.

- Second, robo-taxis don't cause traffic jams. It doesn't take a collision to cause a traffic jam. In Chapter 2, we discussed how so-called "phantom traffic jams" can arise for no reason at all. One irregular braking incident can have a ripple effect—altering the driving behavior of all proximate vehicles. In heavy traffic, this ripple can become a tsunami—affecting every car down the line and ultimately causing gridlock. Humans are not very good at anticipating future navigation decisions nor at responding to sudden changes in velocity. Given the complexities of high-speed road travel and the many external distractions that drivers must contend with, it's surprising that US traffic fatalities are as low as they are—currently hovering around 37,000 deaths per year. Remaining attentive and gauging the velocity of neighboring cars is a tiresome exercise for human drivers. Fortunately, if there's one thing that computers are good at, it's monitoring extremely boring situations. The sensor array on a self-driving car can perform such tasks all day long without ever losing interest.

- Finally, in the more distant autonomous future, robo-taxis won't need to stop at intersections. We mentioned MIT's Senseable City lab in previous chapters. Here, researchers have proposed a "slot-based" system for managing the flow of incoming autonomous cars at intersections. If a robo-taxi's onboard computer can someday be linked to the controls of every traffic-light in the country, then (in theory) cars may someday take you from Point A to Point B without a single tap of the brake pedal. Such an innovation would save countless man-hours and greatly amplify the productive output of the nation.

We don't know the extent to which ground transportation will improve once all of our cars are equipped with the above-described bag of tricks. Several attempts have been made to model the behavior of autonomous vehicles in urban environments. MIT's Senseable City Lab has estimated that ridesharing vehicles could someday handle our entire urban transportation workload with 80% fewer cars. Tony Seba arrived at a similar calculation. In 2014, a team of researchers working with the University of Texas modeled autonomous ridesharing vehicle behavior using conventional traffic data from Austin.

They were able to show that the travel requirements of Austin residents could potentially be fulfilled by a convoy of robo-taxis that was a tenth of the size of a conventional car fleet. Speaking at the Automated Vehicle Symposium in San Francisco, contributing researcher Dan Fagnant said:

> **[We were able to attain a] 10-to-1 household to vehicle substitution. [Meaning that] you can take about ten ... personal household vehicles and serve the same number of trips — with [just one] fully autonomous shared vehicle... [So we're] cutting the total vehicle miles traveled, ... saving the environment, and hopefully cutting congestion as well.**

Achieving such impressive stats in a computer simulation is the easy part. But our clever algorithms are often humbled by the contingencies of reality. Real roads are full of surprises—like bounding boulders, tumbling tumbleweeds, drunk drivers, and wayward wildlife. Such anomalies may spook early versions of self-driving software—perhaps causing these cars to revert to timorous driving strategies that may annoy proximate human drivers. We'll be talking more about the importance of maintaining this man-machine détente in future chapters. For now, while such speedbumps are sure to hinder the development of this technology, there is little doubt that the world of self-driving cars will *someday* be more efficient than the human-driven alternative. If one robo-taxi can do the work of ten conventional automobiles then this achievement would lead to substantially less congestion on our roads.

II. People will run fewer errands in the future

While standing atop a freeway overpass and watching the interminable rows of cars dashing below, one is often struck with a single thought:

"Where are all these people going?"

Thanks to the 2009 National Household Travel Survey by the US Department of Transportation (DOT), we can answer this question.

According to their data, most Americans have four main reasons to take to the streets:

1. To commute to or from work
2. To run errands
3. To go to school or church
4. For recreational or social outings

Average Miles Traveled Per Day

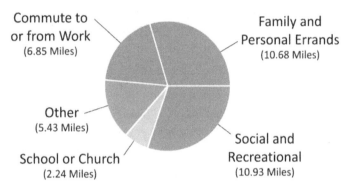

Figure 63 - Average miles traveled per day—arranged by task type. (Source: 2009 National Household Travel Survey by the US Department of Transportation Federal Highway Administration.)

In the previous section, we described how the use of robo-taxis will reduce urban congestion and make all of our trips more efficient. But once autonomous home delivery vehicles are deployed, it's possible that we won't have many road trips to make anyway. This becomes evident if we drill down into the Department of Transportation's four categories and examine the types of errands that people are actually logging. For example, in the "Family and Personal Errands" section we find the following tasks:

1. "Pick up someone"
2. "Drop off someone"
3. "Transport someone"
4. "Buy gas"
5. "Take-and-wait" (The DOT defines a "take-and-wait" as a trip that requires a driver to take a passenger to a destination, wait for them to finish their business, and then drive them back home.)

As every parent knows, such tasks are a major time sink—possibly comprising the majority of modern-day childrearing efforts. However, in the coming autonomous future, parents will no longer be doing *any* of these tasks. Family members will simply use their phones to summon robo-taxis as needed. In a world where nobody drives, nobody can ask anyone else for a ride.

Continuing along in the same DOT section we spot a couple retail trips:

1. "Shopping or errands"
2. "Buy goods: groceries / clothing / hardware store"

As described earlier in this chapter, future consumers will not need to shop at brick-and-mortar stores—instead electing to have most of their purchases delivered to their homes via autonomous couriers. Thus, many of the trips in these categories will be eliminated as the chore of shopping continues to migrate to the digital domain.

Finally, in the "Social and Recreational" section we spot two trips that are solely devoted to retrieving meals:

1. "Go to pick up or eat a meal"
2. "Go to eat, coffee, ice cream, or snacks"

As described in Chapter 5, the majority of our meals will someday come to us via autonomous home delivery. We won't be queuing up in front of fast-food drive-through windows nor retrieving Styrofoam containers from across restaurant take-out counters. Most of our future meals will just drive themselves to our homes. So the number of road trips in this category will be diminished as well.

We don't know how many of our daily errands will eventually be outsourced to the machines. As an exercise, take a moment to reflect upon the utility of your own commonly driven routes. What did you do today?

- Did you get gas?
- Did you pick your kids up at school?
- Did you buy dinner at a drive-through window?

The "future you" won't be doing any of these things. As the autonomous infrastructure evolves, the number of human-performed driving chores will continue to diminish. The majority of errands that require the movement of parcels or people will be completed via a tap on an iPhone. Such innovations will lead to momentous societal changes. Perhaps someday, when *all* of life's trivialities are fulfilled by machines, the very meaning of the word "errand" will change—just as it's done before. (In Old English, the word "ærende" commonly referred to the delivery of a message.) When the age of autonomy is truly upon us, there may still be many vehicles on the road, but the vast majority of them will not be carrying a single soul.

The Road of Today

The Road of Tomorrow

Figure 64 - Instead of human passengers, the roads of the future may primarily service autonomous delivery vehicles.

III. Autonomous car trips will be facilitated by more efficient roads and urban layouts

Thus far, we have examined how autonomous vehicles will use the road more efficiently. Now, let's consider how these vehicles may prompt a redesign of the road itself.

Consider the following images:

Typical 8 lane freeway with shoulders and a median

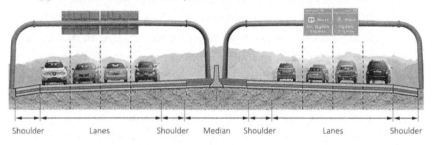

Shoulder   Lanes   Shoulder   Median   Shoulder   Lanes   Shoulder

4 lane freeway for autonomous vehicles

Figure 65 - Thanks to the efficiency and safety of autonomous vehicles, future roads may be constructed with fewer lanes and perhaps without inside shoulders or road signage.

This diagram depicts a typical eight-lane highway as well as a rendering of what the same highway might look like in the age of autonomy. Here, you'll notice that we have diminished the necessity of four key structures: shoulders, medians, lanes, and road signage. Let's discuss each of these alterations now:

1. **Shoulders** run adjacent to the lanes of a motorway. On US freeways, they often lie on both sides of the road—at the periphery and in front of the median (as shown in our image). Their purpose is to allow drivers to escape from the flow of traffic in the event of an accident or engine failure. Road shoulders are an expensive addition (taking up the space of two additional lanes), yet they are rarely used. They are only necessary when car accidents occur. But recall that around 94% of accidents are incited by human error. So when humans are eliminated from the driving equation, then road shoulders may be rendered superfluous. Unlike human drivers, autonomous vehicles will always be in constant communication with

each other—listening for reports of upcoming traffic anomalies and recalculating their route on the fly. If an autonomous car stalls on the road, the other cars will automatically adjust their speed and driving behavior to compensate for the impediment. Like a river winding around a boulder, autonomously guided traffic will flow around the errant car until it can be safely removed from the road. Achieving such regimental coordination between multiple vehicles is only possible if *every* car on the road is an autonomous car. So these conveniences will only be possible in areas that go completely driverless. In such places, the need to construct internal shoulders may be eliminated.

2. **Medians** comprise the area of land that divides two opposing lanes of traffic. Some medians are merely used to beautify roadways with landscaping—possibly featuring grass, trees, flowers, shrubs, or stones. But in high-traffic areas, their primary purpose is to prevent accidents on one side of the freeway from spilling into the other side—resulting in catastrophic head-on collisions. Medians take up about as much space as one or two additional lanes. They may also be host to concrete barriers (sometimes called "Jersey barriers") which further act to prevent crossover traffic. In our diagram, notice that we have removed most of the median area from our highway— only maintaining enough space to accommodate a single concrete divider. Remember, because auto collisions will be so rare in the age of autonomy, we need not devote so much land to the purpose of merely keeping these cars separated. They'll keep an eye on each other as they drive.

3. **Lanes** are the conduits on which cars navigate to their destination. US freeways often boast four lanes in each direction—sometimes several more. Such high lane counts are necessary when humans are at the wheel. But, as described throughout this book, autonomous vehicles should be able to utilize the available road space more efficiently. Their suite of onboard sensors will allow these cars to speed down the road in tight formations, all while constantly optimizing their route and watching for any disturbances in traffic flow. Additionally, most commuters will be in ridesharing taxis— which carry at least four people per vehicle instead of just one. With

such improvements, large multi-lane motorways may no longer be needed. Notice in our diagram that the eight-lane highway has been whittled down to a mere two lanes on each side.

4. **Road signage** looms above our motorways and dots our landscapes with strange symbology—totem poles constructed by those who trekked upon this earth in an era of great darkness—before the invention of GPS. Unfortunately, we can't get rid of these signs yet. Some people still use them to get around. So we must continue to tolerate this litter of multi-colored navigational arrows, road condition alerts, and warnings about future street-sweeping hours— at least for a while longer. But in our lower diagram, notice that the road of the future does not contain any signage at all. Future passengers will not be navigating—their cars will manage each route for them. Signage will no longer be of any use. If a passenger is curious about his current whereabouts, then he can simply refer to the GPS app on his cellphone. Like the pillar-saints in the days of the Byzantine Empire, all of our signs will eventually make their way down from atop their poles. Some will find a new home in museums, others will end up in the collectibles category on eBay. Most will be recycled for their aluminum. Eliminating the need to construct, install, and clean the millions of road signs that line each highway will save billions of taxpayer dollars each year.

In light of the above four observations, we are hopeful that future highways will cast a smaller footprint on the earth, take fewer resources to build, and yet (thanks to autonomous technology) will be the most efficient passenger routes ever constructed.

Thinking beyond the grid

Since antiquity, urban planners have divided their city blocks with crisscrossing perpendiculars constituting the geometry of a basic grid. In such arrangements, intersecting roads meet at right angles—forming a hatchwork of interconnected streets.

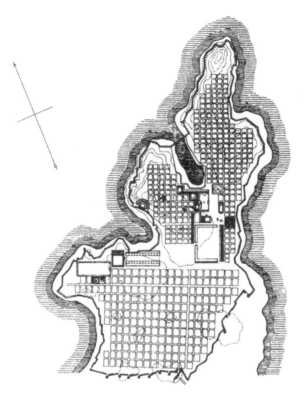

Figure 66 - Grid plan for the ancient Greek city of Miletus (circa 470 BC).

Such grid plans have served us well for a long time.

- They use the ground efficiently, residential blocks are easy to layout with rudimentary tools and a simple coordinate system can be employed for assigning addresses.
- Rapid travel through the city is possible given that such grids allow citizens to swiftly walk across at a zigzagging diagonal—without being cast down interminable corridors.
- Wayfinding does not require advanced navigational skills—thanks to the logical placement of each successive city block.

Simple street grids work well in modern-day urban environments too. But given the particular driving characteristics of autonomous vehicles, it may be time for us to reexamine the utility of the grid.

Currently, human-driven car trips end when we stable our vehicles in a driveway, garage, or parking lot. But robo-taxis and autonomous couriers have no use for such respites. Their only imperative is to drop off or pick up their cargo (of parcels and passengers) in the quickest manner possible. The bottleneck in this enterprise will be primarily determined by the degree to which future city planners are able to remove any impeding choke points in the city's conduits. Unfortunately, our cities currently contain a lot of chokepoints. Namely: culdesacs, intersections, and 90-degree turns.

Culdesacs and intersections break the forward momentum of delivery vehicles—forcing the car to perform inefficient U-turns or halt its forward movement entirely. The 90-degree turns that hatch our rectangular city blocks are fine for pedestrians—who have an average walking speed of 3.1 miles-per-hour. But (as known by anyone who has ever made a hard left in a car), 90-degree turns are not so conducive to a quick and comfortable ride in a fast-moving vehicle.

Tomorrow's urban planners may come to find that the traditional city grid is not the most efficient layout for the age of autonomy. Instead, the roads that interconnect our future dwellings will be free of culdesacs, they will discourage intersections, and they may be molded into the shape of curvaceous splines.

Figure 67 - Instead of using a typical city grid, future cities might someday utilize roads in the shape of graceful splines—which are more conducive to the rapid drop off and pick up of parcels and passengers. (Source: Design by Matthew Spremulli for MIT's Center for Advanced Urbanism "Future of Suburbia" Biennial Research Project 2014-2016.)

When personal car ownership is rare, when autonomous vehicles are the primary means of transportation, and when goods are expected to arrive at doorsteps within thirty minutes of purchase, then cities can facilitate this enterprise by constructing such frictionless conduits. By eliminating the grid and altering the road geometry in this fashion, we can curtail the need for vehicles to engage in excessive bouts of turning or stopping—thus multiplying the efficiency by which autonomous ridesharing and home delivery vehicles traverse residential housing tracts. Using such roads should result in shorter delivery times for courier vehicles and a more comfortable ride for robo-taxi passengers.

IV. Future buildings will be designed with rapid drop-off and pick-up zones.

We have concentrated our discussion thus far on the method and medium by which people of the future will travel to their destinations. But what's going to happen when they actually get there? How will people exit their vehicles in the future?

Currently, our road trips terminate when we park our cars upon arrival. Most of us are familiar with how costly this exercise can be. Owning an automobile means forever being yoked to the burden of parking. For each destination that you arrive at, your first action-step is to find a suitable paddock in which to pasture your steed. In heavily-trafficked corrals, drivers must laboriously instigate a brute-force search for an open parking space—all whilst praying that the newly discovered slot is within walking distance to their intended address.

Over the last 75 years, dozens of research projects have attempted to interview drivers in congested downtown traffic and determine where they were headed. Donald Shoup (professor of Urban Planning at UCLA) reviewed sixteen traffic studies conducted between 1927 and 2001. He found that (on average) 30% of cars cruising in congested urban centers were merely searching for a place to park. A more recent 2007 study found that an astonishing 45% of drivers interviewed on the streets of Brooklyn were just circling the block—looking for curbside parking.

Whatever the numbers are in your town, we can all agree that the chore of parking sucks. Thus far, the default solution to this pickle has been to demand that builders devote ever more acres to parking lots. When space can't be carved out of the ground, developers are forced to build vertically—stacking cars into enormous multi-story concrete structures. Such monoliths can be cost-prohibitive to new developments—further stalling the construction of new homes and contributing to the housing crisis.

Thankfully, as we move to a ridesharing paradigm, many have discovered that getting dropped off at a building's front entrance is much more convenient than terminating each car trip with a trek through the *house of horrors* that is the modern parking lot. Ridesharing vehicles don't park—so they don't need a parking space. What they *do* need is an easy way to drop off and pick up passengers in front of residences and commercial buildings. Unfortunately, most of our structures are not currently well designed for this exchange.

All too commonly, when instigating a passenger drop-off, we find ourselves in an anxious situation in which we must halt the flow of traffic behind us, and jockey our autos into position along a yellow-colored curbside. This maneuver must be executed while other drivers are honking and screaming at us from over our left shoulder. For a demonstration of this cacophony, try observing your local elementary school at 8:00 a.m. on a Monday morning. Take note of the school's drop-off process and watch as hordes of frantic parents compete for pole position and then jettison their cargo of children to the curb.

If we are to fully actualize the efficiencies of the autonomous age then not only must our *cars* be redesigned, but our drop-off processes must be restructured as well. Future buildings will eventually be equipped, not with parking lots, but with upgraded drop-off and pick-up zones—capable of handling several passenger exchanges per minute. In Chapter 4, we talked about how future homes may be equipped with a street-facing vestibule designed to accept parcels and people from incoming autonomous vehicles.

Figure 68 - The street-facing vestibule in this home is designed to receive incoming autonomous deliveries of parcels and people.

Such designs could work for some residences. But what about commercial buildings, hotels, apartment complexes, shopping malls, and retail districts? How will these structures receive incoming robo-taxis?

Thankfully, many contemporary architects are well aware of the need for such design alterations. Some are already constructing new developments with ridesharing drop-off and pick-up zones in mind. Andy Cohen's team at the Los Angeles architectural firm Gensler, has designed the new "Epic" building in Hollywood to come equipped with a "coach lobby" — a drop-off zone that sits adjacent to the building's interior reception area—thus facilitating a car-to-building passenger transfer. Instead of demanding that ridesharing vehicles plop their passengers onto the building's outside curb, this "coach lobby" sits inside the building itself—easing the transformation of residents to riders.

Figure 69 - A rendering of the "coach lobby" to be built in Gensler's new "Epic" building in Hollywood, California. The driveway runs inside the main lobby entrance so that ridesharing vehicles can rapidly drop off and pick up passengers from inside the facility.

Lamenting how intrusive and uglifying the automobile has been to urban developments, and acknowledging the need for improvements in curbside pick-up, Cohen remarked:

> **The buildings we're designing today...have to be designed for the era of driverless cars with...major drop-off and pick-up areas...in front... The [building's lobby will become a] great portal—for people coming or leaving a building... Think about all the real estate frontage of...our streets... [We can take that back—reclaim it for]** *people-space,* **for** *green-space,* **for** *amenity-space,* **restaurants, or cafes... This is our big opportunity (as architects, designers, and urban planners) to take our city streets back.**

Automobiles have been dictating the design of our cities for far too long—pushing their weight around and molding our urban environments to suit their own needs, not ours. It is indeed time to "take our city streets back." To reclaim the beautiful avenues of our cities and design them, not for cars, but for people—like we used to do.

A 2010 study by researchers at the Department of Civil and Environmental Engineering at Berkeley, estimated that parking facilities represent somewhere between 0.5% and 12% of total estimated lifecycle energy consumption and greenhouse emissions, and 24% to 81% of other air pollutants. Whatever their actual environmental impact may be, there is little doubt that these vast asphalt deserts alter the local ecology wherever they are laid. It would be nice to see an America where our one billion parking spaces can be either utilized for more productive means or simply allowed to return to a more natural state. Perhaps we have been living atop this flatland for so long that we have failed to realize the multitudinous dimensions to be revealed after these asphalt shrouds are finally lifted.

Will we have Urban Density or Urban Sprawl?

Getting the populace to switch from personal car ownership to ridesharing robo-taxis should allow for more efficient urban land use—simultaneously liberating existing parcels for development (parking lots can be turned into homes) and increasing accessibility to the untrodden territory at the city's periphery (people will be able to reside far outside the city center). Concomitant with these benefits should come shorter commute times and a decrease in traffic congestion. If the coming mechanical marvels succeed in delivering upon such audacious claims, then, for the first time in history, the common man will be able to engage in daily interaction at the city center, while residing hundreds of miles away from it—all without any exorbitant commuting costs.

It's worth taking a moment to consider how people will respond if such an efficient transportation infrastructure is finally perfected. How will these innovations alter our modes of living?

- Will people view the introduction of autonomous long-range passenger vehicles as an opportunity to flee the gritty blight of the city?
- Or will the widespread use of autonomous delivery vehicles allow people to burrow ever deeper into their urban hermitages?

Let's consider both of these outcomes now:

Outcome 1: People choose to get out of Dodge.

The ease by which future autonomous vehicles will traverse long distances may incite an increase in sprawl and an exodus to the suburbs and exurbs. It could be that some types of dwellings are more reflective of human biological imperatives than others. We can perhaps get a glimpse into these primitive inklings by examining the types of homes that people search for online.

83% of Zillow.com users devote their attention to considering single-family detached houses—rather than townhouses, condos, duplexes, or mobile homes. Perhaps the stereotypical "little house with the white picket fence" is reflective of our primal aspirations—to be part of a community, and yet, to also carve out a niche for ourselves. Thus, it may be the case that, when autonomous vehicles finally offer the urbanites an opportunity to vault the concrete barricades of the city, most of them will elect to do just that. The land that sits at the periphery of our urban zones may soon be inundated with city slickers in pursuit of a new homestead.

Outcome 2: People pile into the cities.

Given the coming improvements in autonomous people-moving and home delivery, the congestion of future cites should be more tolerable for urbanites. As described in this chapter, city conduits *feel* crowded because of the sheer number of people and cars that must compete for space on the street. However, many of the tasks that currently require people to take to the streets (e.g. transporting children, hunting for parking, picking up dinner, etc.) will all someday be performed by machines. Additionally, the acres of parking lots on which conventional autos currently sit, are soon to be made obsolete—plopped onto the real estate market and open to residential developers. Thus, it is likely that population density will continue to increase in our cities. Yet, thanks to the emerging autonomous infrastructure, this coming swell may be adequately accommodated. Thanks to this technology, urbanites should live quite comfortably in the future, despite increased urban density.

Outcome 3: People do both

Undoubtedly, *both* of the above-described outcomes will occur in varying degrees. Some nine-to-five office workers will be happy to utilize autonomous technology to flee to the countryside when quitting time comes. Others will be glad to bid adieu to the country bumpkins each evening and return to their swanky apartment buildings. If the coming autonomous infrastructure can simultaneously handle the fastidious demands of urbanites and the ebb and flow of super commuter traffic, then *both* of these subspecies may harmoniously thrive; both modes of living should be viable in the city of the future. As Dan Fagnant told The Economist in 2015:

By liberating space wasted on parking, autonomous vehicles could allow more people to live in city centres; but [these

cars] would *also* make it easier for workers to live farther out… [There may be a] simultaneous densification of cities, and expansion of the exurbs.

Building smarter instead of building more

In this chapter, we have tried to expose the many costs that a civilization must bear to maintain the apparatus of daily commuting. Merely moving a person from their home, to their job, and back to their home again is a logistically complex endeavor—requiring an enormous infrastructure that is currently fraught with inefficiencies. The high cost of housing is a result of these inefficiencies. To host conventional automobiles, millions of acres must be devoted to redundant lanes, garages, and parking lots. To keep the wheels of commerce spinning, we must cut up the land into graded plains—dicing it up into grids, scalping the local peaks with excavating equipment, and uprooting the indigenous flora that charm the earth.

Near my town in California, State Route 73 is a good example of such profligate cultivation.

- The road doesn't cater to any residential zones. Instead, it cuts directly through the uninhabited (and formerly untouched) acres of the Laguna Coast Wilderness Park and the San Joaquin Hills.
- 68 bridges had to be constructed to traverse the many canyons that once slivered across the ungraded California terrain.
- 32 million cubic yards had to be excavated to flatten this land for asphalt.
- Though the road is only a modest 17.7 miles long, its construction cost taxpayers $800 million.

In glancing at the following map of Route 73, you'll notice that it starts and ends at the same northbound road that it runs parallel to. A driver who decides to take the Route 73 off-ramp will pop out at precisely the same place as a driver who does not.

Figure 70 - California State Route 73 runs parallel to the existing Interstate 5 and 405 highways.

So why was Route 73 built?

It was built to reduce traffic congestion along Interstates 5 and 405 in South Orange County—potentially allowing drivers to avoid the crunch and save a few minutes by taking an alternate route. In other words, Route 73 only exists because of the inadequacies of present-day modes of travel. The many shortcomings of conventional automobiles are not only detrimental to the pocketbooks of their human drivers, but also to the ground on which they tread—millions of acres must be devoted to nothing more than black asphalt.

Let us hope that, thanks to the efficiencies to be gained in the coming age of autonomous transport, roads like Route 73 will not need to exist. The natural California wilderness need not be host to asphalt plains and concrete bridges. And the local taxpayers need not be prompted to dole out millions of dollars for a toll-road that lies adjacent to an existing eight-lane interstate highway.

Thus far, our go-to strategy to deal with America's increasing number of cars has been to keep increasing the number of roads.

- We construct redundant highways that run parallel to existing ones.

- We stack our freeways atop each other using expensive vertical corridors.
- And we encapsulate old motorways with ever-broadening streets.

But we can't keep adding lanes forever; roads are not Russian dolls capable of unbounded accretion. Thankfully, there is hope that the transportation innovations of the coming autonomous age will finally allow us to halt this infinite regress.

Ch. 8: Social Disruption

"The Machine Stops"

Some time ago, *The Oxford and Cambridge Review* published a 12,000-word short story about a future dystopian civilization. Several common sci-fi tropes make an appearance in the narrative. For example:

- Each citizen of this world lives in a one-room apartment in a towering residential complex.

- Dating, marriage, and sex have fallen out of custom. Instead, people prefer to spend their days engaged in trivial intellectual pursuits—chatting online, listening to music, or watching academic lectures via streaming video feeds.

- There is no need to work in this future age. Autonomous machines cater to every whim of the populace—diligently fetching food, drinks, and clothing as needed.

- International airline travel is free but it is rarely used. Globalization has run its course—each city on the planet is indistinguishable from the next.

- Everyone in this world has thousands of "friends" with whom they communicate online. But they haven't actually met any of them in person. Instead, they prefer to do almost all of their communication via a video conferencing device called an "optic plate"—on which they can push a few buttons and call up anyone on earth.

- Since there is no need (and little desire) to venture outdoors and mingle with their fellow man, many suffer from social anxiety and they have trouble interacting with the physical world outside of their tiny apartments.

Does any of this sound familiar?

Perhaps you've heard this story before. You might be thinking that this "future forecast" is all too obvious and cliché. Some elements of the story might be reminiscent of your own life or the life of someone you know. But a little historical context is needed here:

- Zoom launched in 2011.
- The iPhone launched in 2007.
- YouTube launched in 2005.
- Facebook launched in 2004.
- Skype launched in 2003.
- The first website launched in 1991.
- The Apollo Moon Landing was in 1969.
- The first satellite launched in 1957.
- The first television station opened in 1928.
- The first transatlantic flight was completed in 1927.
- The first transcontinental telephone call was placed in 1915.

And *this* story, "The Machine Stops" by English author E. M. Forster, was published (over a century ago) in 1909.

Figure 71 - Portrait of E. M. Forster (born 1879) by the English painter Dora Carrington.

The Death of "Going Out"

Many science fiction pieces warn of the dangers that await us atop the ladder of scientific progress. Often, the characters in the story use technology to achieve a utopian civilization, only to quickly devolve into a state of decadence or torpor. For our purposes here, *The Machine Stops* makes for a useful parable because it describes a society that has thoroughly mastered the challenge of automation—albeit at great cost to the human psyche.

In the novel, we are introduced to Vashti. Like every other citizen in her world, she lives alone—in a solitary apartment adorned with buttons by which she summons *The Machine* to bring anything she desires.

[Vashti's] room was studded with electric buttons... There were...buttons to call for food, for music, for clothing. There was the hot-bath button... There was the cold-bath button. There was the button that produced literature, and there were of course the buttons by which she communicated with her friends.

The room, though it contained nothing, was in touch with all that she cared for in the world.

The Machine Stops by E. M. Forster (1909)

By examining Vashti's world, perhaps we'll reveal the trajectory of our own—particularly in reference to the unseen consequences that may occur following the adoption of an autonomous infrastructure. Below, we'll contrast our world to that of Vashti's in three domains: *dining*, *media*, and *socializing*.

#1: Dining in the Age of Automation

In *The Machine Stops*, people don't go out to shop or buy groceries. Nor do they ever go to restaurants. Instead, any dish they desire is delivered to their apartment by *The Machine*. Consequently, each meal is consumed alone—in front of their computer terminal, in the seclusion of their tiny apartment.

What about our world?

With the growth of online meal ordering apps (like UberEATS, DoorDash, and GrubHub), many people are choosing to stay home and dine-in. In the 2018 UBS Investment Bank study cited earlier, it was revealed that millennials were three times more likely than their parents to order from meal delivery apps. Forecasting the eventual commoditization of the American "home-cooked meal," they stated:

> **Thanks to robot chefs, drones, delivery droids, and shared dark kitchens, the cost of delivered meals could be falling sharply...The [cooking] expertise [that consumers] currently [use] in-house...could potentially be rendered [obsolete]... Or [consumer] expertise might shrink to [merely] preparing breakfast or cups of tea—much like sewing has arguably shrunk to basic clothing repairs carried out at home.**

The automation of home food delivery will undoubtedly increase the number of people who elect to outsource the chore of cooking. Food-delivery apps now fall within the top 40 most-downloaded cellphone apps in most major markets. So, at least for single people, it is likely that "dining alone" (on meals that are summoned via the tap of an app), will someday be the norm.

#2: Media in the Age of Automation

In *The Machine Stops*, Vashti spent every hour of her waking life consuming internet media—attending lectures, listening to music, and chatting with her online "friends." According to the book, she had "several thousand" of them.

What about our world?

We need not go into much detail about the many corollaries between Vashti's online activities and those of our own. Over the last decade, social media has insinuated itself into every aspect of our lives—from dating and relationships to religion and politics. Conversations about the pernicious effects of getting "too much screen time" are peppered into every newsfeed. A 2016 study by the consumer research group Taylor Nelson Sofres, found that the average millennial spends an incredible three hours-per-day on a mobile device—almost double the screen time of their Gen-X predecessors. If trends continue, it's likely that media-consumption rates of successive generations will eventually become competitive with Vashti's stats.

#3: Socializing in the Age of Automation

In *The Machine Stops*, citizens rarely leave their apartments and they never communicate in-person anymore—instead preferring to keep all of their relationships online. Though Vashti has several thousand friends she has never actually met any of them in the flesh. There is no dating, sex, or marriage in her world. Human contact is so rare, that even a momentary touch can cause great anxiety. Such an incident happened when Vashti was helped on board an airship by a flight attendant.

...the attendant of the airship, perhaps owing to her exceptional duties, had grown a little out of the common. She had often to address passengers with direct speech, and this had given her a certain roughness and originality of manner. When Vashti swerved away from the sunbeams with a cry, [the attendant] behaved barbarically—she put out her hand to steady her. "How dare you!" exclaimed [Vashti]. "You forget yourself!" The [attendant] was confused and apologized for not having let her fall. People never touched

214

one another. The custom had become obsolete, owing to The Machine.

What about our world?

Most of us need only scroll through their own Facebook contact list to see a lengthy roster of "friends" — most of whom we don't really know very well. Perusing one's digital list of "followers" on Twitter and Instagram tends to yield similar results. Is new media turning us all into misanthropes? Are we to devolve into a nation of keyboard warriors—waxing poetic on the internet at all hours of the night, but afraid of our shadows during the day?

The American psychologist Jean Twenge has spent her career studying changing social norms for young people. Along with Heejung Park, she wrote an influential paper titled *The Decline in Adult Activities Among U.S. Adolescents* which analyzed the survey responses of 8.3 million US teenagers from 1976 to 2016. Her paper indicates exactly what its title implies — "adult activities" are declining among US adolescents. Kids are less likely to get a driver's license, get a job, get a date, or try alcohol.

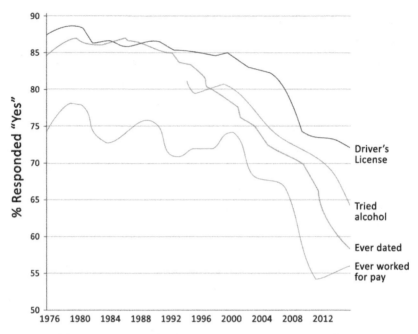

Figure 72 - The percentage of US 12th graders who have 1) a driver's license, 2) have tried alcohol, 3) have gone on a date, and 4) have worked for pay. (Source: Twenge and Park's The Decline in Adult Activities Among US Adolescents, 1976-2016.)

Millennials do still go on the occasional date. But *online* dating has become the norm—with 19% of couples reporting that they met through a dating website or app. This tops the 17% who met "through a friend," and the 15% who met "at a bar or restaurant." According to the 2019 GSS, the portion of Americans (ages 18 to 29) reporting "no sexual relations in the past year" has more than doubled since 2008.

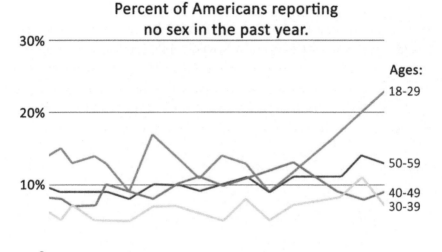

Figure 73 - The number of young people (ages 18 to 29) who reported having *no sex in the past year* has doubled since 2008. (Source: General Social Survey (GSS) 2019 by the National Opinion Research Center.)

But even when they do manage to get a date and get laid, fewer young people are deciding to tie the knot. Marriage rates in the US peaked after World War 2 and have been on the decline for decades—recently reaching an all-time low in 2020 following the coronavirus pandemic.

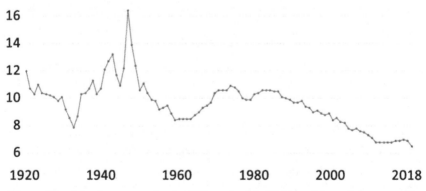

Figure 74 - Marriage rates in the US peaked in 1946 and have been steadily declining since the 1980s. (Source: Data compiled by *Our World in Data* using: Carter et al. (2006) for the 1920-1995, the US Census Bureau (2007) for 1996-2004, and the CDC for 2005-2018.)

Though there are more opportunities to socialize than ever before, people are choosing to eschew human contact. The 2018 American Time Use Survey by the U.S. Bureau of Labor Statistics reported that citizens now spend an average of only forty-one minutes-per-day socializing with others. That's down nine percent from the previous decade. Even younger people are socializing less. As Twenge and Park's paper noted, a steadily declining number of them are motivated to venture outside of the home without a parent.

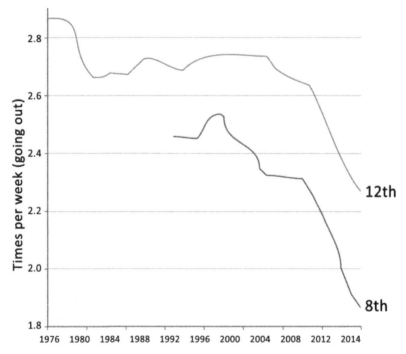

Figure 75 - The number of times per week that US children leave the house without their parents. (Source: Twenge and Park's The Decline in Adult Activities Among US Adolescents, 1976-2016.)

The rise of the hipster curmudgeon (and his associated antisocial proclivities), could result in irreparable economic repercussions. Some call it "The Death of Going Out." But the phenomenon also goes by many industry-specific names:

- "The death of the movie theater"
- "The death of the shopping mall"

- "The death of the arcade"
- "The death of the night club"

Such businesses have failed to attract millennials to the same degree that they attracted their parents. The advent of social media, online dating, and the recent flux in courtship mores, have surely negatively affected the prospects for singles hoping to get together in such environments. According to a controversial 2017 Economist/YouGov poll, 17% of 18 to 29-year-old Americans believe that the act of a man "inviting a woman out for a drink" either "always or usually" constitutes sexual harassment.

It may be the case that the death of these traditional entertainment providers is the inevitable result of the newfound ease by which we digitally summon in-house fun. In the past:

- The night club didn't have to compete with Tinder.
- The arcade didn't have to compete with Nintendo.
- The shopping mall didn't have to compete with Amazon.
- The movie theater didn't have to compete with your 60-inch plasma television set.

All of these emergent competitors have undoubtedly coaxed many potential consumers into canceling their dinner reservations—instead opting for an evening at home to "Netflix and chill." This change of plans is reflected in the poor performance of movie ticket sales—which peaked in 2002 and have been declining ever since. In 2017, theaters sold about as many tickets as they sold in 1995—even though the US population gained 59 million people.

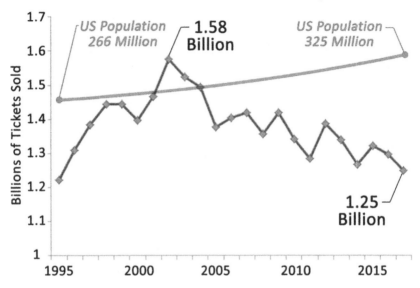

Figure 76 - Despite an increasing US population, movie theater ticket sales peaked in 2002 and have been declining ever since. (Source: Plot originally rendered by Wolfstreet.com using ticket sales data provided by The-Numbers.com.)

We just don't go to the movies as much as we used to. We also don't go to the mall, or to the club, or to the bar, or to the bowling alley anymore. One wonders how long it will be before the consumption of amusement via public social gatherings is considered too risky (or too expensive, or dangerous, or boring, or déclassé).

- Are such conventional forms of entertainment domed to fade into obsolesce entirely?
- Will we even bother to leave our homes at all in the future?
- Is this the death of "going out?"

"Going out sucks anyway"

The marvels of the coming autonomous infrastructure will make it easy for us to manifest the reclusive angels of our nature. When man eventually succeeds in constructing machines that are capable of fulfilling all home deliveries in thirty minutes or less, he may run out of reasons to leave his house. Such conveniences could result in a nation of shut-ins—each of us residing in solitary dwellings, hypnotized by cellphone screens, and filling every waking moment with computer games, porn, and cat videos. Perhaps, for many, such a hermit-friendly culture is exactly the type of society that they long to be a part of.

In a 2016 article titled "Millennials Have Discovered 'Going Out' Sucks," Vice News writer Harry Cheadle concluded:

Fuck going out… It's expensive, it's crowded, it smells bad, the bands are usually terrible, the clubs are usually worse. You know what's great? Sitting around and watching TV… You get to wear comfortable clothes, summon whatever food you want via phone and eat it with your hands… [When compared to most] of the humans who have ever lived, this generation's typical "night in" represents an impossible pinnacle of luxury… a glass of wine, a roof overhead, and a screen that can show you anything you can imagine.

Cheadle's rant (accompanied with a smattering of recent dismal social trends), seemed to have struck a chord with many. In the least, such observations rang true enough to inspire a series of amusingly titled web articles.

Figure 77 - The horrors of a "night on the town" are discussed in these posts from four popular news websites.

Present-day innovations *already* allow us to eschew the primal act of venturing out onto the gritty streets of the real world and (instead) to hunker down in our solitary media-saturated hermitages. Given that this lifestyle seems to appeal to so many of us in present-day America, we can infer that such modes of living will be even more alluring after autonomous courier technology has been perfected. After all, if you can summon food, wine, and entertainment to your couch (sans delivery fees), then why bother leaving it?

If we are to follow along with this logical progression, then we may come to the same conclusion as E.M. Forster's protagonist:

She had studied the civilization that had immediately preceded her own... [In that civilization, they had] mistaken the functions of the system... [They] had used it for *bringing people to things*, instead of for *bringing things to people*.

What a mystifying display of perspicacity...

I think Forster's conjecture will turn out to be correct. All of our science fiction films got it wrong. Hollywood would have us believe that the future is one in which high-technology enables people to jet around the world and sample the amenities of exotic lands. But it is more likely that the opposite will be true. Instead of fleets of flying cars *"bringing people to things,"* your actual future will mostly consist of autonomous machines *"bringing things to people."*

Figure 78 - Sci-fi shows like *The Jetsons* often feature flying cars "bringing people to things." But future transportation technology will mostly consist of legions of automatons "bringing things to people."

Given that current technology has already succeeded in turning so many of us into *Diogenes with an iPhone*, it seems likely that future upgrades will only hasten this inversion. Ironically, we may eventually discover that autonomous vehicles are *much* better at keeping us apart than they are at bringing us together.

The pretense to socialize

The ability to reside in a metropolis—without the need to interact with a single soul—is a feat that is only made possible given recent technological advances. In days of old:

- The town square provided shopping venues
- The office watercolor provided a cold drink
- The nightclub dance floor provided girls
- The church provided God
- The fire provided warmth

But everyone knows that the product itself was not the sole impetus for utilizing these facilities. Each one acted as a pretense to socialize, to gossip, to find a wife, or to bond with your fellow countrymen. So what will happen to the nation, when all of these products are exclusively consumed at home? Once we thoroughly master the art of "bringing things to people," it may be the case that people will stop bringing themselves to each other.

Already, most of us find ourselves tasked with manufacturing a "reason to go out." Something prods us to bear the exorbitant cost of such ventures, even when we understand that they can be attained at home for a much-reduced cost.

- The price of cuisine served in a Michelin 3-star restaurant, doesn't necessarily correlate with its taste or nutritional value.
- Alcohol can be mixed and poured for a fraction of the bar's fee. In double-blind taste tests, even trained sommeliers can't distinguish expensive wine from cheap wine (or even red wine from white wine).
- Almost every new movie is streamable shortly after its initial theater release.
- And, after a couple of drinks, cranking up your home stereo is about as good as any audio experience you might pay for at your local live music venue.

How long will society keep up this charade?

As we improve the efficiency by which goods and services are delivered to our homes, we may find ourselves tasked with a comical conundrum—in which we will forever be on the hunt for an increasingly more sophisticated pretense to go outside.

Figure 79 - Patrons at Cinema Pathé in Switzerland can pay $50 to recline in their new "bed seats"—designed to replicate the experience of watching movies at home for free.

Figure 80 - Patrons at the "Cereal Killer Café" in London can pay £8 for the experience of eating a £1 bowl of breakfast cereal. (Photo by Londo Mollari.)

The Great Irony...

It has never been easier to "find something to do." And yet, if we are to believe the U.S. Bureau of Labor Statistics, most of us don't like doing very much. Futurology has long promised us a world of flight and effortless travel. But such achievements may not have the same appeal to future generations. It is possible that our wanderlust will be quelled just as the apparatus needed to facilitate it becomes available to all. This is what E.M. Forster predicted:

Few travelled in these days, for, thanks to the advance of science, the earth was exactly alike all over. Rapid intercourse, for which the previous civilization had hoped so much, had ended by defeating itself.

What was the good of going to Peking when it was just like Shrewsbury?

Why return to Shrewsbury when it would all be like Peking?

Men seldom moved their bodies; all unrest was concentrated in the soul. The air-ship service was a relic from the former age. It was kept up, because it was easier to keep it up than to stop it or to diminish it, but it now far exceeded the wants of the population.

The Machine Stops by E. M. Forster (1909)

One wonders if this is to be our fate.

- Could it be that future transportation innovations become so efficient, that they succeeded only in eliminating the need for their own existence?

- Will the effortless transport of people, parcels, and media (or "intercourse" as Forster calls it in this passage) ultimately diminish the differences between regions so thoroughly, that tourism is rendered superfluous?

- Will the world finally become "flat"—as the journalist Thomas Friedman would say?

Coinciding with the development of autonomous vehicles will come progressively more impressive advances in communication and media technologies. The long-held promise of the "virtual office" has only recently started to be delivered upon. In 2017, the Census Bureau reported that 5.2% of workers in the US worked at home—up from 3.3% in 2000. Given the lockdown stipulations of the 2020 COVID-19 pandemic, we expect this value to surely increase in the years to come. Additionally, the virtual reality industry (after decades of failures) has finally managed to produce a commercially successful product line—the Oculus (acquired by Facebook in 2014 for two billion dollars). With this sort of financial backing, new VR innovations are sure to be on their way. What will happen when the virtual world starts to pass for the real world?

- Why fly around the globe to visit the Forbidden City of Peking, when you can see it via 3D virtual reality goggles?

- Why order Fish & Chips in Shrewsbury, when your local kitchen center will prepare a dish with a similar recipe and deliver it to your front door via autonomous courier vehicles?

The great irony about autonomous vehicles is this: By the time there exists a massive transportation infrastructure—able to effortlessly whisk passengers to any location on Earth—many people may discover that they don't particularly want to go anywhere. If current social trends continue, the activities that fill our lives may start to look more like those of E.M. Forster's protagonist.

She made the room dark and slept; she awoke and made the room light; she ate and exchanged ideas with her friends, and listened to music and attended lectures; she made the room dark and slept.

Above her, beneath her, and around her, The Machine hummed eternally; she did not notice the noise, for she had been born with it in her ears.

The Machine Stops by E. M. Forster (1909)

Ch. 9: Economic Disruption

> We are being afflicted with a new disease of which some readers
> may not yet have heard the name, but of which they will hear a
> great deal in the years to come—namely "technological
> unemployment." This means unemployment due to our
> discovery of means of economising the *use of labour,*
> outrunning the pace at which we can find new *uses for labour.*
>
> – Economic Possibilities for our Grandchildren
> by John Maynard Keynes (1930)

Garbutt, New York

In a 2016 piece for National Review, Kevin Williamson painted a bleak
portrait of Mr. and Mrs. America. In the background of his montage, the
country's many blue-collar towns lie in a state of ruinous decay. Many of
these communities have failed to thrive in today's global economic reality.
Citing the tiny New York hamlet of Garbutt, Williamson wrote:

> The truth about these dysfunctional, downscale communities
> is that they deserve to die. Economically, they are negative
> assets. Morally, they are indefensible. Forget all your cheap
> theatrical Bruce Springsteen crap. Forget your sanctimony
> about struggling Rust Belt factory towns and your conspiracy
> theories about the wily Orientals stealing our jobs...The
> white American underclass is in thrall to a vicious, selfish
> culture whose main products are misery and used heroin
> needles. Donald Trump's speeches make them feel good. So
> does OxyContin. What they need isn't analgesics, literal or

231

> **political. They need real opportunity, which means that they
> need real change, which means that they need U-Haul. If you
> want to live, get out of Garbutt.**

Over the last thirty years, many manufacturers have left New York State. Xerox, GM, Kodak, Corning, and Bausch have all moved overseas—taking thousands of factory jobs with them. Given that most of these companies will not be returning to the US, perhaps Kevin Williamson is right—maybe small towns like Garbutt do "deserve to die." Perhaps, if the residents of Garbutt seek economic opportunity, they should rent a U-Haul truck, say goodbye to grandma, and relocate to a more economically enlightened area. Williamson writes:

> **Yes, young men of Garbutt—get off your asses and go find a
> job: You're a four-hour bus ride away from the gas fields of
> Pennsylvania.**

For now, simply moving their "asses" might work. But, in the coming world of autonomous vehicles, would such a deracination improve their plight?

Garbutt lies 18 miles south of Rochester, 61 miles east of Buffalo, and 94 miles west of Syracuse. Getting a job in any of these three cities would be possible, but would require a commute time that most residents wouldn't find desirable. The average American commute is 16 miles in length—putting some parts of Rochester in reach but not Buffalo or Syracuse. In the following image, we've drawn a 16-mile radius around Garbutt's three neighboring cities.

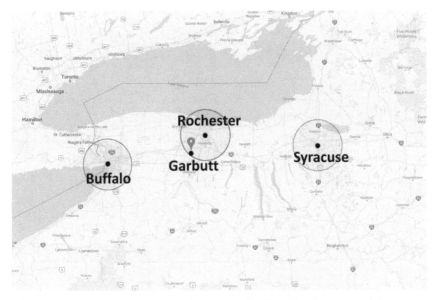

Figure 81 - Garbutt, New York lies between Buffalo, Rochester, and Syracuse. A 16-mile radius (the average American Commute distance) is drawn around each city.

Now, let's consider what our maximum commute distance might look like, in a world of autonomous vehicles.

- This is a world in which nobody owns a car. Instead, a robo-taxi picks you up at your house in the morning and drives you to work while you recline.
- This is a world where vehicles on the interstate (guided by the sure hand of A.I.) can casually reach speeds of around one hundred miles-per-hour.
- It's a world without parking lot labyrinths.
- A world without traffic jams.

In such a world, traversing a longer commute distance may not require a longer commute time. And it definitely won't require any mental effort on the part of the passenger—he can sleep on the way. So, a much longer commute would be tolerable—perhaps of, say, 80-miles. Below, we have redrawn our map such that our three cities now depict this new radius.

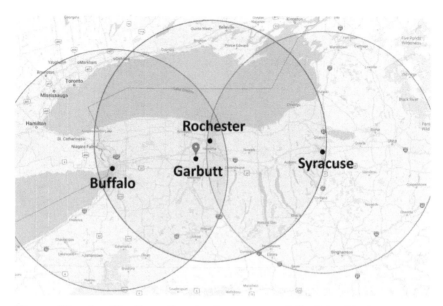

Figure 82 - An 80-mile radius is shown around Buffalo, Rochester, and Syracuse. Autonomous vehicles may make long-distance commutes more feasible.

Autonomous vehicles will allow small satellite towns (like Garbutt) to supply larger (more economically viable) cities with labor. Thus, to expand their occupational horizons, the future citizens of Garbutt won't necessarily "need a U-Haul" as Williamson suggests. They just need driverless cars. These vehicles would give them access to jobs throughout upstate New York—thus solving their economic woes.

So that's good, right?

Maybe.

An autonomous infrastructure may initially seem like a winning solution for all. Until you realize that expanding the radius of job opportunities for Garbutt, necessarily means expanding it for everyone else too. The labor force of Garbutt will be competing with the labor force of neighboring Scottsville, Belcoda, Mumford, and a hundred other small towns you've never heard of. And therein lies the rub: in a world of effortless travel, the employee's proximity to the job is no longer such an economic advantage. Currently, there is great value in minimizing the distance between the worker's home and the job location itself. Workers must reside within a manageable radius around their place of employ. Autonomous vehicles will allow us to multiply this radius by a factor of four (possibly much more). Thus, the available supply of labor—for any given job in the nation—will inflate.

When this happens, something as simple as a "U-Haul" will not be enough to save the citizens of Garbutt. That is to say, merely relocating to a more economically enriched area might not increase their job prospects. Residents will be competing with the same expanded labor supply (inflated by the use of autonomous vehicles) whether they decide to leave Garbutt or not.

In contemplating the future, our concern is not that there won't be enough jobs in America's forgotten little towns (like Garbutt). Instead, there may not be enough jobs for anyone, anywhere.

The Four Horsemen of the (Economic) Apocalypse

In keeping with the gloomy mood of this chapter, let's invoke some apocalyptic imagery. Viktor Vasnetsov's 1887 painting of the "Four Horsemen of the Apocalypse" appears on the next page. (Each rider has been slightly modified with Photoshop by the author.)

1. Drivers
2. Food Workers
3. Retail
4. Everyone Else

THE FOUR HORSEMEN OF THE ECONOMIC APOCALYPSE

We'll use these horsemen to denote the four sections of this chapter—concentrating on the effect that the coming autonomous revolution will have on four job sectors:

1. Drivers (Transportation and Material Moving employs 10 million Americans.)
2. Food workers (Food Preparation and Serving employs 13 million Americans.)
3. Retail (Sales and Retail Stores employ 15 million Americans.)
4. Everyone else.

Note that the order in which these occupations appear does not necessarily describe the order in which these industries will be disrupted. Each trade will go through periods of death and reinvention—possibly several times before they are entirely appropriated by the machines. In this chapter, we will attempt to describe some of the events that will lead up to this acquisition.

The 1st Horseman of the Apocalypse: "Drivers"

Trucking is the most common job in 29 states. The average trucker is male, 46 years old, and has a high school education—with little or no college. With salaries hovering around $49,000 annually, interstate trucking is one of the most lucrative jobs for blue-collar workers. Trucking employs over 3.5 million Americans. According to the BLS, about two million of these jobs are classified as "heavy and tractor-trailer truck drivers." These are the men paid to haul 80,000 pounds of cargo down the highway—a time-consuming and costly undertaking.

The first technology company that manages to teach a robot how to drive a truck will be in a position to tap into billions of freight dollars. Morgan Stanley estimated that automating the industry would save $168 billion per year. ($35 billion in saved fuel, $70 billion in reduced labor costs, $36 billion due to fewer accidents, and $27 billion thanks to increased asset utilization.) Given such appealing numbers, one might wonder why it's taking so long to get autonomous trucks on the road. But such forecasting is complicated; primarily because the rate at which trucking routes will be usurped by autonomous vehicles is dependent upon the diligence of our A.I. programmers and the willingness of our legislators to risk allowing these vehicles to mingle on public roads—a potentially litigious undertaking. However, several existing prototypes have already shown promise.

In 2019, Volvo's autonomous semi-trailer truck (code-named "Vera") got her first job transporting containers at APM Terminals in Gothenburg, Sweden—Volvo's hometown.

Figure 83 - Volvo's autonomous semi-trailer truck prototype—code-named "Vera."

Other prototypes have already driven across thousands of miles of US highway. But we are still several years away from widespread implementation—in large part due to the fact that there is *much* more to "being a truck driver" than just "driving the truck." The skills required to drive down a stretch of open highway are far removed from the skills needed to, say, negotiate a big rig through downtown traffic and then detach its trailer at a dirt construction site. In 2018, the American Center for Mobility (ACM) examined the threat that automation posed to the trucking industry. They confidently concluded:

The transition to automated driving in the trucking industry is ... anticipated to be gradual. Moreover, in the foreseeable future, automated vehicles could supplement, rather than substitute vehicle operators ... allowing freight transportation and other delivery service companies to address an existing labor shortage. In the coming decade, automated vehicles will not significantly, if at all, impact truck driving jobs.

Given the many complexities of urban navigation, most trucking jobs require a *human touch*. However, practiced finesse is usually only necessary at key moments in the trip. During the vast majority of the ground-transport process, not much is going on. The cargo must simply be hurried down the highway as the driver negotiates his rig across long stretches of uneventful road. While autonomous trucking technology might not immediately take over the entire trucking enterprise, it is during such monotonous stanzas that its utility will be most appreciated.

This is how the process might work:

1. A human driver would drive his rig to a meeting point outside of the city—where he would then release his trailer to a waiting autonomous truck.
2. This autonomous truck would then haul the trailer for the longest stretch of the route—typically across interstate highways.
3. When the autonomous rig finally approaches its destination, it could stop at a waypoint outside of town and release its trailer to a second human driver—who would then drive the rig into the city.

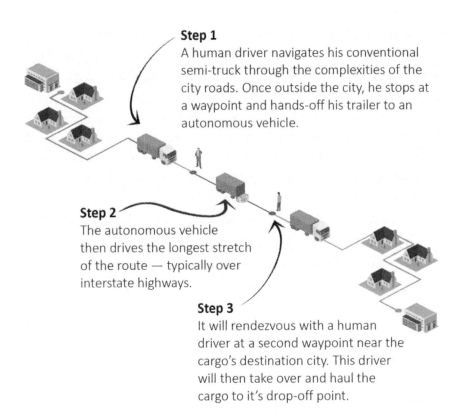

Step 1
A human driver navigates his conventional semi-truck through the complexities of the city roads. Once outside the city, he stops at a waypoint and hands-off his trailer to an autonomous vehicle.

Step 2
The autonomous vehicle then drives the longest stretch of the route — typically over interstate highways.

Step 3
It will rendezvous with a human driver at a second waypoint near the cargo's destination city. This driver will then take over and haul the cargo to it's drop-off point.

Utilizing this method, a human driver is only required for the most tedious parts of the journey—the points where trailers must be hitched, hands must be shook, invoices must be signed, and safety protocols must be adhered to. Once our human driver has managed the "hard part" of truck driving, then he can simply offload his trailer to an autonomous vehicle—one that is quite capable of driving across thousands of miles of featureless (and very boring) terrain.

If this method becomes the norm, then there will still be plenty of room for human drivers in the industry—at least for the coming decade. Autonomous vehicles may initially only succeed in appropriating some of the more routine trucking tasks. A human touch may be required for years to come. However, the method by which the human drivers "take the wheel" might change.

"Taking the Wheel" via Remote Vehicle Operation

In May of 2019, the Swedish transport company (Einride) began daily freight deliveries for a DB Schenker logistics facility in Jönköping, Sweden. Their autonomous truck (the T-Pod) makes short-haul trips—moving cargo between the warehouse and the terminal. Though its current top-speed is limited to 5 km-per-hour, the implementation is significant because Einride is one of only a handful of companies to be issued a permit to operate an autonomous vehicle on a public road. Speaking to Reuters, Einride CEO Robert Falck said,

---

This public road permit is a major milestone [and it's a] step to commercializing autonomous technology on roads.

Additionally, the Einride truck is unique because it allows a driver to take control remotely. A human operator can drive the truck using a PC terminal—which may be located thousands of miles away from the vehicle itself.

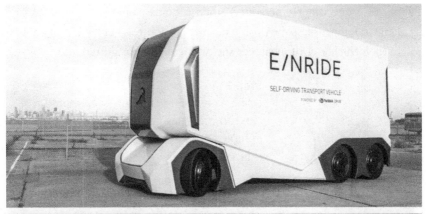

Figure 84 - Thanks to a 5G network connection, the Einride autonomous vehicle can be remotely piloted. Here, a convention attendee in Barcelona, Spain uses a terminal to drive a prototype vehicle around a test track in Boras, Sweden.

This feature is noteworthy because it could (in theory) allow for a method by which future transportation companies can bypass any logistical or legal complexities that would prevent autonomous vehicles from driving in some locales. In this schema, drivers could login to an autonomous rig whenever a human touch is required. For example, if a particular municipality doesn't allow fully-autonomous vehicles on their roads, a human driver could take control of the rig remotely—only to relinquish control back to the A.I. software as soon as the vehicles passed the nearest county line. Or, if an alternate contingency arises (like a snowstorm, traffic accident, mechanical failure, or some other emergency), then a human operator will always be available to login, assess the problem, and take control—at least long enough to get the vehicle out of whatever sticky situation it has gotten into.

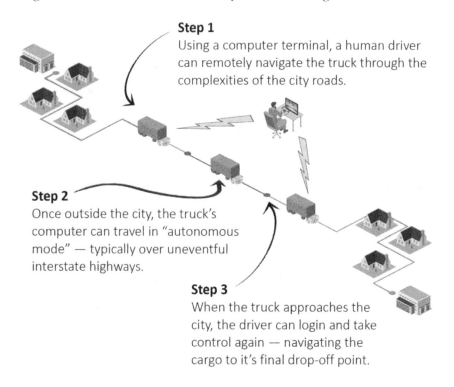

Step 1
Using a computer terminal, a human driver can remotely navigate the truck through the complexities of the city roads.

Step 2
Once outside the city, the truck's computer can travel in "autonomous mode" — typically over uneventful interstate highways.

Step 3
When the truck approaches the city, the driver can login and take control again — navigating the cargo to it's final drop-off point.

With this method, trucking companies need only employ human drivers to execute the most logistically sensitive driving maneuvers. But, in the least, there will still be some room for human employment. As Einride's CEO was quick to note, this schema doesn't eliminate the driver job entirely:

Instead of sitting inside a truck, [the driver is] operating the system itself. Instead of removing the [driver's job], we're giving him a completely new way of working.

Perhaps transitional solutions like this will become the most prevalent form of freight transport in the early days of the autonomous revolution. Autonomous semi-trucks won't take every trucking job in one fell sweep. Instead, this emergent technology might initially serve to merely complement existing trucking jobs—rather than replace them outright. Developing a transportation infrastructure that is capable of the entirely "touchless" movement of freight (i.e. able to haul cargo without requiring a single human interaction) may not be logistically or legally possible for a few more decades. The same ACM study (cited above) contends that job displacement "will likely not take place until the latter half of the 2020s." This might be welcomed news for a truck driver who is nearing the end of his career. But such a brief time horizon probably won't appease the anxieties of the 21-year-old kid who just got his commercial driver's license last month. At *some* point in this person's career, he'll be racing his truck alongside an *autonomous* truck—one that is *far* more efficient than he'll ever be. (Recall that human truckers are only permitted to drive a total of 11 hours per day.) While *The Trucking Apocalypse* may not happen tomorrow, let us not doubt that it will happen someday.

What about the other driver jobs?

Thus far in this section, we have focused on freight transport occupations. But what about the other drivers—particularly those catering to residential delivery like:

- The Package Delivery Driver (FedEx, UPS, or DHL)
- The Food Delivery Driver (UberEATS, DoorDash, or GrubHub)
- The Mailman
- The Pizza Man

As we discussed in Chapter 4, these types of driver jobs are the most difficult to automate because they require an additional step—curb-to-doorstep delivery. Autonomous vehicles are very good at obeying traffic laws and moving large containers down major highways. But, as they near their destination address, it's difficult for them to make sense of suburbia's many inefficiencies—e.g. the faded address signs, the lack of parking, the barking dogs, the picket fences, the porch stairs, and the locked gates.

Such *last-mile obstacles* are currently only circumnavigated via human finesse. For the time being, the jobs of our delivery drivers—who are naturally able to interact with the world at this phenomenal level—will be safe. Their replacement by autonomous vehicles can only happen if the last-mile of home delivery is sufficiently mechanized—i.e. when our sidewalks and mailboxes are designed to be a bit more machine-friendly. To facilitate the wheels of automation, future residences must be equipped with an apparatus (a dropbox as described in Chapter 4) to receive incoming deliveries. Thus minimizing the number of unpredictable obstacles that would hinder the transfer of parcels from the street to the home. When such an infrastructure is complete, all delivery vehicles will be autonomous vehicles. And all "driver jobs" will be in jeopardy. Aside from the occupations discussed thus far, here are a few others that are destined to go the way of the dodo:

- The Chauffeur
- The Public Bus Driver
- The School Bus Driver
- The Shuttle Bus Driver
- The Auto Transport Driver
- The Tow Truck Driver
- The Armored Car Driver
- The Newspaper Delivery Driver
- The Ice Cream Man
- The Valet Parking Attendant

Millions of existing occupations are heavily reliant upon a man's ability to direct some type of vehicle down the road. Most of these jobs will not survive the autonomous revolution. Morgan Stanley's 168 billion dollar estimate (cited above) only covers the first stage of autonomous deployment—mostly in semi-trucks. But once autonomous technology is commoditized—once self-driving software is included in every newly-produced car on the road—then the actual cost of this disruption will be well into the *trillions*.

The 2nd Horseman of the Apocalypse: "Food Workers"

When robo-taxis finally replace personal car ownership, America's 200 billion dollar fast-food industry may be initially impacted the most. Such businesses are heavily dependent upon *location*—serving consumables to a regional clientele via brick-and-mortar stores. Given their short wait times and affordable prices, people typically frequent fast-food restaurants when they need a quick bite—often while en route or during the evening commute. Consequently, some franchises conduct as many as 70% of their transactions via the drive-through window. Because so many locations are required to keep the burgers in reach of the drivers, the USA is home to over 247,000 fast-food restaurants. Multiple locations are often tightly clustered together. In sticking with the upstate New York example that began this chapter, we've compiled a list of eleven major fast-food chains currently operating in the Rochester area. Here, there are:

- 20 McDonald's
- 19 Subway Sandwich Shops
- 14 Burger Kings
- 14 Wendy's
- 13 Taco Bells
- 5 Arby's
- 5 Domino's Pizzas
- 5 Chipotles
- 5 KFC's
- 3 Popeye's Fried Chickens

- 2 Sonic Burgers

This comes to a total of 105 stores. It takes a battalion of food handlers to satisfy the chicken and burger demands of the good citizens of Monroe County. If we assume that each of these 105 locations employs 15 people, then this brings our total number of fast-food workers to 1,575.

To consume this food product, the meal has to somehow make its way from the grill to the customer's mouth. There are only two ways to make that happen: Either the food must travel to the customer or the customer must travel to the food. The coming autonomous infrastructure will alter the way that *both* of these options are executed.

- If our future customer wants the food to come to him, then he need only place an order via cellphone and request a home delivery. In Chapter 5, we described how meals will be purchased online and prepared in large facilities ("autonomous kitchen centers") which will house dozens of restaurants including fast-food brands. Once his burger is prepared, it will be placed in a container, loaded into an autonomous courier, raced across town, and delivered to his door in minutes.

- Now if our customer insists on dining *inside* a chain location, then he'll simply pull out his cellphone, summon a robo-taxi to his home, ride to the restaurant, and eat his meal there.

At first glance, these modifications to the conventional food-ordering process may not sound particularly disruptive to the restaurant industry. But note that, in the age of autonomy, the distance that lies between the patron and the kitchen is no longer such a constraining variable. Thanks to autonomous couriers, servicing the entire Rochester area may only require a dozen autonomous kitchen centers—not the 105 fast-food chain locations that are currently operating there. Because the cost to travel *one mile* via robo-taxi will eventually be only marginally different than the cost to travel *ten miles*, there will be almost no benefit in selecting one franchise location over another. Robo-taxis will put an expanded number of potential dining locations in reach.

For example, if our patron has a craving for a Big Mac, the robo-taxi app can prompt him to select one of the 20 McDonald's locations currently operating around Rochester. So which one will he choose?

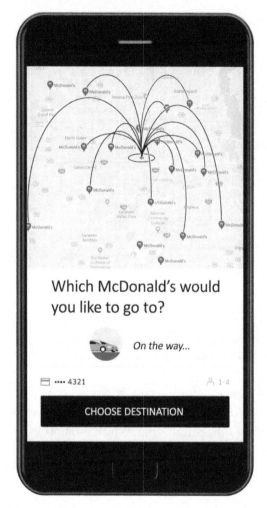

Figure 85 - When the cost to reach any given McDonald's restaurant is negligibly different, then multiple chain locations will not be needed in the area.

In the age of autonomous travel, our patron's dining options are virtually indistinguishable. Note that in the franchise business model, restaurants do not differentiate themselves by offering a "unique dining experience." Chain stores (by their very design) are all supposed to be exactly the same.

"Proximity to the customer" is the only way that one McDonald's can distinguish itself from the other 19 McDonald's locations operating around Rochester.

As depicted in the following map, note that each McDonald's is an average of only 2.2 miles away from its nearest neighboring McDonald's.

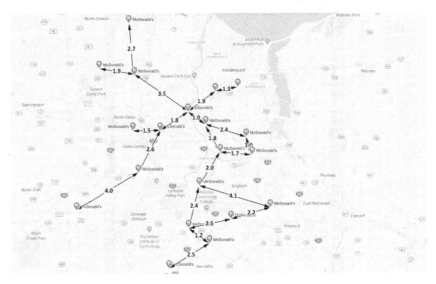

Figure 86 - Each McDonald's in Rochester is an average of only 2.2 miles away from its nearest neighboring McDonald's.

Once the autonomous infrastructure is finally complete, traversing an additional distance of 2.2 miles might only add a few seconds and a few pennies to the cost of the patron's trip. Which prompts the question:

In the age of autonomous vehicles, why would we need 20 McDonald's restaurants around Rochester?

When the driving time and cost required to move your body from your home to any given regional McDonald's location is inconsequentially different, then there will be no benefit in operating more than one McDonald's in the area. Thus, we conclude that the vast majority of fast-food chain locations (in any given city) will close. Only a handful of demographically strategic locations will remain. And they will serve an ever-broadening customer radius—that will continue to expand as autonomous technology becomes more efficient.

Instead of 20 McDonald's franchises operating around Rochester, there may eventually be only one or two. Of course, this goes for Burger King, Wendy's, Taco Bell, and every other chain in the area. The 1,575-person labor force that is currently needed to run these places may be whittled down to a couple hundred employees—thus leaving the remaining 1,300 workers without a job. This won't just happen in Rochester of course. Every cluster of chain restaurants in America may eventually experience a similar consolidation event. As far as I know, theoretical economic geography doesn't have a name for this phenomenon. So, I'll make up my own name and call it the "Chain Store Apocalypse."

Chain Store Apocalypse

/CHān stôr ə'päkə‚lips/
noun

The onset of market saturation following the end of personal car ownership and the adoption of autonomous robo-taxis as the primary means of transportation. When the consumer's cost (in time and fare) to travel to any given chain location is negligibly different, then there is no benefit in the operation of more than one chain location in the region. Thus, the introduction of autonomous vehicle technology will cause the majority of chain store locations to fail.

* Note: *Along with restaurant chains, this also applies to retail chains and supermarket chains.*

In the future, fulfilling the cheeseburger demands of an entire county may only require a couple large McDonald's mega-kitchens—not dozens of little kitchens positioned at high-traffic points. If individual chain restaurants are to survive, they will need to differentiate themselves utilizing a lure other than *location.*

Figure 87 - These two full-service McDonald's locations in downtown Los Angeles are only 1900 feet apart. Given the navigational complexities of LA's roads, both businesses have no trouble luring customers. But if effortless autonomous travel becomes a reality, then there will be no benefit in operating such proximate facilities.

It is already the case that most conventional diners and bars offer more than just *food*. They function as a watering hole—providing both sustenance and social comforts to the local wildlife. To keep the customers coming back, chain locations might consider fostering a regional clientele and offering ancillary amenities. They'll have to do *something* different. Because, with the dawning of the autonomous age, mere *proximity* will no longer be a viable selling proposition.

The 3rd Horseman of the Apocalypse: "Retail"

The in-store shopping experience

Given that every nightly business report invariably features *some* sort of news about a recent success won by Amazon, eBay, or Alibaba, one might assume that online sales account for a large slice of the retail pie. But the majority of purchases are still transacted the old fashioned way—across a checkout counter of a brick-and-mortar store. If we exclude on-site spending (like at restaurants, bars, auto dealerships, and gas stations), then E-commerce only makes up 16% of total retail sales.

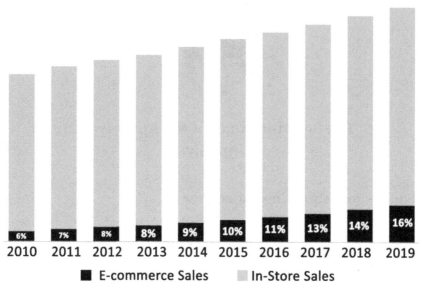

Figure 88 - E-commerce sales continue to grow, but still only make up 16% of total retail sales. (Source: Data originally compiled by Digital Commerce 360 using of U.S. Department of Commerce data- Monthly Retail Trade report from February 2020.)

Of the top ten US retailers (Amazon, Walmart, Kroger, Costco, Home Depot, CVS, Walgreens, Target, Lowe's, and Albertsons) each one except Amazon still conducts the majority of their business in the offline world. With the exception of a few niche markets, people prefer to shop in a physical store for the majority of their daily purchases. This is especially true for things like food, groceries, candy, shoes, clothing, makeup, toiletries, haircare products, medication, home décor, lumber, and hardware.

Though online sales continue to grow, online retailers have yet to convince customers to purchase the bulk of their home essentials via internet. This failure has been attributed to the many inefficiencies of the digital domain. Aside from all of the familiar logistical difficulties that make e-commerce cost-prohibitive for many household products, we'll list a few other deficiencies:

- Each online transaction comes with additional delivery fees or wait times—products often take several days to arrive at your front porch.

- Brick-and-mortar stores offer an experience that that can't be provided via a laptop screen. Customers have the opportunity to conduct a tactile inspection of the product before purchase. And if the customer has any questions, there's the potential for an interpersonal customer service encounter.

- There's also a social element to shopping. For some, walking through a mall is akin to walking through a museum. The store's wares may be arranged in a meticulously coordinated display—often featuring vivid colors and accented lighting. Such a shopping experience is difficult to replicate digitally.

Figure 89 - Williams-Sonoma is known for its elaborate in-store merchandise displays. Though their products are also sold via their website, the shopping experience is much different in their brick-and-mortar stores.

Perhaps someday, our VR headsets will succeed in simulating reality well enough to allow us to peruse products via a virtual stroll through a virtual store. But until that day comes, people will have to (somehow) physically move their bodies to the brick-and-mortar location to fully actualize the *complete shopping experience*. We currently accomplish the feat by getting into a car and driving ourselves to the store's parking lot. But how will *future* customers go shopping?

Going shopping in the future

Around the Rochester area, we see many grocery and retail chains including:

- 19 Wegmans Supermarkets
- 16 Tops Friendly Markets
- 16 CVS Pharmacies
- 11 Walgreens
- 9 Walmart's
- 6 Targets
- 5 Home Depots
- 3 Lowes

Like the fast-food model discussed in the previous section, these chain stores are designed to offer the same shopping experience in every locale. "Proximity to the customer" is the only way that one CVS Pharmacy can differentiate itself from the 15 other CVS Pharmacy locations in the Rochester area. In the below map, you'll notice that each CVS is an average of only 3.1 miles away from its nearest neighboring CVS.

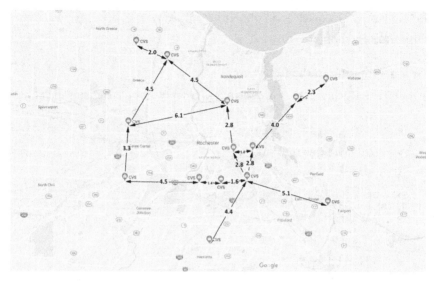

Figure 90 - This map shows each CVS Pharmacy location around Rochester, New York. Each CVS is an average of only 3.1 miles away from its nearest neighboring CVS.

In the coming world of autonomous vehicles, traversing a distance of 3.1 miles might only add a few seconds and a few pennies to your journey. Which prompts the question:

In the age of autonomous vehicles, why would we need 16 CVS Pharmacy locations around Rochester?

When the driving time and cost required to move a customer from his home to any CVS location is inconsequentially different, then there will be no benefit in the operation of more than one or two CVS locations in the area. Like the fast-food restaurants we discussed in the previous section, brick-and-mortar retail stores (like CVS) will also be susceptible to the *Chain Store Apocalypse*. Only a couple demographically strategic locations will be needed to serve the citizens of Rochester. And these stores will serve an ever-broadening customer radius—one that will continue to expand as autonomous technology becomes ever more efficient. If this is to be the fate of all such retail chains, then the labor force needed to run each franchise will substantially decrease.

Online Sales and the Retail Apocalypse

In 2016, the National Center for Education Statistics reported that 81.9% of US households have internet access and 89.3% of households own at least one computer or cellphone. The recent ubiquity of these devices has enabled consumers to make more e-commerce purchases. Online sales continue to rise each year—reaching $187.2 billion in the fourth quarter of 2019.

Ever since the birth of the internet, people have been predicting the death of brick-and-mortar retail stores. The term "Retail Apocalypse" was popularized in 2017 following the announcements of a series of bankruptcies and closures by hundreds of classic American brands: Payless, Radio Shack, The Limited, Gamestop, J.C. Penney, Kmart, Macy's, and Sears.

Figure 91 - Founded in 1893, Sears was the largest retailer in the United States until 1990. Sears filed for Chapter 11 bankruptcy in October of 2018. (Photo by Daniel Case.)

Curiously, the ongoing apocalypse came at a time of national economic growth—with rising GDP and increased consumer retail spending.

So where was all this money going?

An over-supply of shopping malls, an outdated customer experience, and changes in consumer buying behavior are all in part to blame for the apocalypse. But the most cited factor is a shift in consumer spending—away from traditional department stores and toward emerging e-commerce retailers.

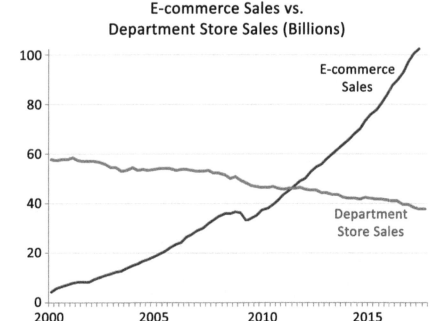

Figure 92 - This 2019 plot by Wolfstreet.com illustrates the decline of department store sales, contrasted with the aggressive rise of e-commerce sales.

But recall that most sales are *not* transacted in department stores. In the previous section, we noted that *online sales* still only account for 16% of *total retail sales*. While this value is better than the meager 6% share of a decade ago, online retailers still have a lot of ground to cover if they're ever going to compete with their brick-and-mortar counterparts.

The public has been persistently reluctant to utilize e-commerce for the majority of their household purchases. This hesitation is primarily a result of the remote nature of e-commerce. In the traditional brick-and-mortar model, the customer becomes the physical steward of his newly purchased product immediately following the chime of the cash register's bell. But when it comes to online sales, the customer won't be able to get his hands on the product until it travels hundreds of miles—from a warehouse to his porch. This trip requires money and time—products usually spend several days in transit; exorbitant shipping fees have always been the thorn in the side of the online consumer. Such burdens add to the transaction cost of every online purchase and make the ten-minute drive to your local Walmart seem all the more appealing.

Amazon Prime

Much of this book has been devoted to describing the many logistical complexities of order fulfillment. Amazon has been working diligently to ease this burden since its inception. Jeff Bezos founded the company in July of 1994—selling books out of the garage of his Washington home—the company's first warehouse. Bezos now oversees an impressive network of over 175 fulfillment centers (occupying over 150 million square feet) throughout North America and Europe. When Amazon Prime launched in 2005, customers benefited from "free" two-day shipping—for a fee of $79 annually. By offering millions of products via a single platform and repackaging product shipping costs into a single flat fee, Amazon helped to alleviate the *pain of paying* and reduced the mental transaction cost that is so typical of online purchases. For many online customers, Amazon Prime was a welcomed addition to the world of e-commerce. According to Consumer Intelligence Research Partners, an astonishing 62% of US households were Amazon Prime members in 2019.

When it comes to product delivery, Amazon initially relied on conventional mail carriers like USPS, UPS, and FedEx. But with the launch of Amazon Flex in 2015, Amazon has embraced the Uber-fication of business. A network of independent contractors (aka "gig workers") are employed to help carry the 3.5 billion packages that Amazon delivers annually. As of 2019, about half of all orders are fulfilled by Amazon's own package delivery network.

Such innovations have made online transactions more appealing for consumers. But companies like Amazon—who aspire to completely mechanize order fulfillment—will someday reach a point where every logistical bottleneck has been conquered except one—their reliance on human packers and human drivers. Amazon already operates over 200,000 warehouse robots. But they also employ over 840,000 humans. Amazon associates and robots currently work together (in harmony no doubt) on the warehouse floor—"a symphony of humans and machines," as Amazon Robotics chief technologist Tye Brady put it.

As evidenced by Amazon's ever-increasing stock price, this relationship seems to be doing quite well at the moment. Human employees still play a critical role in order fulfilment. The human hand is an amazing thing. Mechanical arms don't have the finesse needed to play classical violin nor the dexterity to tape up cardboard boxes full of diapers and Dixie cups.

Or do they?

In 2019, Amazon began testing automated packing machines—capable of creating custom-sized shipping boxes for outgoing customer orders. Reportedly, the packing machines could replace 24 Amazon warehouse jobs each. Matt Novak of Gizmodo noted that Amazon was installing the new machines while simultaneously offering existing employees $10,000 to quit their warehouse job and "create their own startup" — delivering packages for Amazon via their gig worker delivery platform. Novak wrote:

> **...you may be thinking to yourself, "isn't this just an effort by Amazon to...create an army of people who look like they work for Amazon... but who don't actually work for Amazon?" Yes, and the company has been using these kinds of tactics for a very long time.**

Such "employee opportunity programs" may be a harbinger of things to come. As the age of autonomy grows near, existing workers may be prodded off the warehouse floor and perhaps offered gig work or some other transitional position to fill—at least until the construction of their "replacement bot" is complete. Since 2012, Amazon spent around three billion dollars acquiring robotic startups like Kiva Systems, Canvas Technology, Dispatch, and Zoox. Amazon engineers are currently hard at work—teaching their new bots how to pick and package your products without the need for any human involvement in the process.

Given the complexities of order fulfillment, present-day warehouse workers and delivery drivers need not be immediately alarmed. They'll have plenty of work to do—at least for another decade or so. But it's not *this* generation of Amazon employees that we're worried about. It's the next one.

There will come a day when the *entire* order fulfillment process does not involve any human interaction whatsoever. A newly purchased product will be picked by an autonomous warehouse robot, loaded into an autonomous delivery vehicle, and deposited into the autonomous dropbox in front of the customer's home. When companies like Amazon finally achieve their goal (of moving a blender from the warehouse to your front porch in thirty minutes or less), then the retail apocalypse will be complete.

This should give you pause. Because, according to the 2014 US Census, more Americans work in *retail sales* than in any other occupation in the country.

The 4th Horseman of the Apocalypse: "Everyone Else"

In the previous sections, we discussed the collapse of three key job sectors: drivers, food workers, and retail. If you're not currently employed in any of these industries, then you might assume that your occupation will forever be safe from the robot's chopping block. But your confidence may be misplaced. The disruption of these trades will have repercussions that will reverberate throughout every sector of the national economy. As these three dominos fall, so too will many of their adjacent markets.

Take "auto sales" for example. Of the 16,753 franchised car dealerships tracked by the National Automobile Dealers Association (NADA), a total of 17.2 million light-duty vehicles were sold in 2018—accounting for over $1 trillion in new-vehicle sales. On top of this tally, these dealerships fulfilled over 310 million repair orders—with *parts and service* bringing in another $116 billion. 2018 was a great year for car sales and servicing. But the party may be coming to an end.

Currently, about 87% of the US driving-age population has a driver's license—that's around 230 million of us. In 2018, 273 million vehicles were registered in America. But when robo-taxis replace personal car ownership, nobody will be purchasing or servicing automobiles anymore. So there will be no reason for car dealerships to exist. Those glass showrooms featuring red sports cars (dizzily spinning on an oversized Lazy Susan) will disappear. So too will the squad of salespeople who work the floor. As well as every ancillary business that exists solely to provide goods and services to car owners. Consider this brief list:

- Auto supply stores
- Auto transport
- Auto mechanics
- Auto loans
- Car washes
- Car insurance
- Garage door installers
- Paid parking lots
- Gas stations

Most of these businesses will not exist anymore—at least not as we know them now. We're talking about billion-dollar industries that employ millions of people. US vehicle insurance alone accounts for roughly $200 billion in annual revenue. Even the humble *car wash and auto detailing* industry tallies up $12 billion. As self-driving vehicles become the norm, the many businesses that exist solely to cater to the wants of drivers will fade into obsolesce.

A similar fate awaits the industries on the periphery of the other two dominos discussed in this chapter—food workers and retail. Consider the 15 accountants, the 15 lawyers, and the 15 window washers who currently provide services for the 15 independently owned McDonald's franchise locations in your city. As described above, when robo-taxis make effortless travel a reality, then only one or two McDonald's locations may still be needed. The services of the many businesses that supported these soon-to-be-lost chain franchises will no longer be required in their current capacity. Similar repercussions will be felt by all of the ancillary employees who currently cater to these industries as well. And thus the dominos will continue to fall. One after the other, after the other, after the other …

Americans are often warned about the dangers of globalization and the monolithic threat of Asian manufacturing. But for the majority of US workers, this concern is secondary. As the Swedish-German economist Carl Benedikt Frey noted:

> **…most people in America today work in services that are delivered locally. In other words, most Americans are [already] shielded from…globalization. But they aren't shielded from [automated checkout machines] or autonomous cars.**

Because so many US businesses are founded upon the *regional* delivery of goods and services, the prospect that one might "lose your job to a Chinamen" is not a direct threat to most Americans. The Chinese don't live in your city. So they can't wash your car or deliver your pizza. Instead, the biggest threat to the American worker will come (not from the Chinese), but from other moneyed industrialists who will deploy fleets of autonomous vehicles—thus forcing millions of jobs (including your local pizza man's gig) into obsolesce.

Will the machines take our jobs?

For generations, academics, economists, and futurists have all spun cautionary tales about the eventual arrival of the robot worker armies. The threat of "losing our jobs to the machines" is nothing new.

In 1811, the Luddites protested the use of automated textile equipment in English factories. Though many forms of protest were used at the time, a small faction in the group was known to break into mills and destroy stocking frames and other machinery. The Frame-Breaking Act 1812 permitted the death penalty for protestors caught engaging in "machine-breaking." Approximately 60 to 70 Luddites were hanged for their crimes. By 1815 the movement had largely subsided.

Figure 93 - Luddites smashing a mechanized weaving loom in England. (Source: 1844 illustration by The Penny Magazine.)

In 1863, the English iconoclast Samuel Butler penned an article titled "Darwin among the Machines" in which he proclaimed:

Day by day...the machines are gaining ground upon us; day by day we are becoming more subservient to them; more men

> are…devoting [the] energies of their…lives to the development of mechanical life… [The] time will come when the machines will hold the real supremacy over the world and its inhabitants… [No person] of a truly philosophic mind can for a moment question [this]… War to the death should be instantly proclaimed against [the machines]. Every machine of every sort should be destroyed by the well-wisher of his species. Let there be no exceptions made… [Let] us at once go back to the primeval condition of the race.

Butler's warning of a future robot uprising sounded laughable to his readers; some thought that the article was meant as a satirical jibe at Charles Darwin. But in successive works, Butler assured his fans that he wasn't kidding around. Still, it must have been difficult to take Butler seriously in the 19th century. However, as we read his words now (one-fifth of the way through the 21st century), Butler perhaps doesn't sound so wacky.

Speaking 156 years later at the 2019 World Artificial Intelligence Conference in Shanghai, Elon Musk seemed to echo Butler's screed:

> There's…a smaller and smaller corner of intellectual pursuits that humans are better than computers [at]… Every year it gets smaller and smaller. And soon, we will be far surpassed [by computers] in every single way…guaranteed. Or civilization will end. Those are the two possibilities.

The political scientist Charles Murray has written about the threat of technological unemployment for decades. Speaking at the Cato Institute in 2016, Murray reiterated his long-held concerns:

> I am increasingly convinced that…within a matter of decades…the number of jobs that disappear [will be] so great, that we have to start thinking in terms of [a different kind of] economy… The response that I get…is, "You guys have been saying this since the Luddites—that every

technological advancement is going to get rid of jobs. And every time you've been wrong. [Instead, more new] jobs have been created. And to argue 'this time it's different' is a fool's game!" [But] this time it's different. It's not just that we're going to have driverless cars (probably in about 10 years). And [that's a] huge number of jobs that [will] disappear. [But, we're also] going to be carving out millions of white-collar jobs. Because, artificial intelligence—after years of being over-hyped—has finally come of age. And it is now able to do all sorts of things that formally were done by people...with college educations. Pretty smart people—who had to make decisions that a computer couldn't make. And now, the computer can do that... Unless we [face these problems now], we are going to be going down the wrong road. Job programs, retraining, [and] all [these ideas that are typically proposed as solutions to our] workforce problem, just aren't going to work—for reasons that are historically unprecedented...

Despite Murray's warning, it is true that the naysayers of artificial intelligence have been correct to doubt the threat of the coming robot apocalypse. Like nuclear fusion, hyper-intelligent A.I. has been "just ten years away" since the 1970s. And some researchers are much more pessimistic about recent A.I. achievements than Murray. A 2017 report by McKinsey & Company attempted to calculate the number of jobs that could potentially be lost to the forces of automation. They acknowledge that many workforce sectors will be disrupted:

...in about 60 percent of occupations, at least one-third of the constituent activities could be automated, implying substantial workplace transformations and changes for all workers.

But they also contend that the many efficiencies to be gained via the implementation of automation technology could spur on economic growth and might actually *increase* demand for human labor. They write:

> **Even with automation, the demand for work and workers could increase as economies grow... A growing and dynamic economy—in part fueled by technology itself and its contributions to productivity—would create jobs. These jobs would result from growth in current occupations...and the creation of new types of occupations that may not have existed before, as has happened historically. This job growth...could more than offset the jobs lost to automation. [More] occupations will change than will be lost... [People will] increasingly work alongside [the machines of automation].**

So, it is quite possible that these industries will proceed along as usual—only gradually changing over the next several decades.

However, though we may debate about the length of the A.I. fuse, let us not doubt that the fuse has already been lit. If we merely assume that human general intelligence need not necessarily ride on a spiritual ether (an ectoplasm substrate powered by "spooky stuff" as the neurophilosophers call it), then it is virtually guaranteed that man will succeed in creating super intelligent machines (at least someday). But will the mental capacities of these machines overtake our own?

In his 1965 paper "Speculations Concerning the First Ultra-intelligent Machine," the British mathematician I. J. Good wrote a pointed proof of the inevitable supremacy of machine intelligence.

> **Let an ultraintelligent machine be defined as a machine that can far surpass all the intellectual activities of any man however clever. Since the design of machines is one of these intellectual activities, an ultraintelligent machine could**

design even better machines; there would then unquestionably be an 'intelligence explosion,' and the intelligence of man would be left far behind. Thus the first ultraintelligent machine is the last invention that man need ever make...

Given this effortlessly persuasive argument, we are left to contend with the notion that (if granted enough time) the machines will eventually be smart enough to do *all* of our jobs better than we can. And, every single person on the planet will be unemployable someday.

In 2014, the Oxford philosopher Nick Bostrom published a poll asking 500 A.I. researchers to estimate a timeframe for the dawn of General Intelligence. The median estimate yielded a 50% chance of "high-level machine intelligence" (an intelligence that can do most professions as well as a typical person) being developed by 2050. With "superintelligence" following less than 30 years after that.

But even if these estimates are too audacious, that's irrelevant to the conversation at hand. Because it only takes a modicum of intelligence to displace most of the planet's workforce. The vast majority of workers are not engaged in labor that is cognitively demanding. Consider the:

- drivers
- material movers
- agriculture and food prep workers
- cashiers
- factory line workers

This is not to claim that these jobs do not employ any sophisticated laborers; all of these occupations surely require moments of intellectual prowess and nimble-fingered finesse. But, if we were to examine the skill caliber needed at any given moment during the workday of these employees, the action they're engaged in only utilizes a modest degree of muscle strength and a willingness to perform a repetitive task. Not every aspect of their jobs will be so easily automated. But with each successive generation of automatons, ever more human tasks are sure to be delegated to the machines. This appropriation will continue until, one day, there may not be many jobs left for humans to do. To illustrate this point, let's try a thought experiment.

The "My Kid Can Do That" Thought Experiment

Imagine a robot endowed with all the intelligence, dexterity, and muscle strength of a ten-year-old boy. But unlike a human boy, this robot tirelessly follows directions—exactly as they're conveyed to it. Though you may have to explain things slowly and carefully at first, once the bot understood the task at hand then it could work on this task continuously—throughout the day and night and all without a single complaint. Do you think you could find a job for such a robot in your company?

Before answering this question, consider the psychological drivers that compel humans to do work. Getting people to show up at the jobsite requires a financial incentive. When we exchange labor for pay, we not only struggle against the weight of our load, but also struggle against ourselves. A billion neurons scream at the conscious mind—imploring it to "Stop working and take a break!"

Children are particularly difficult to employ because they have yet to see the value in exchanging hours of their lives for green pieces of paper. But adults are able to temporarily mute the demands of the lower mind. If properly incentivized they can coax their bodies into doing work—at least until five o'clock comes around.

Robots are different. They never get bored, hungry, thirsty, sick, jealous, aroused, or cranky. Robots never go on strike nor do they ever demand a pay raise. They never sue you for sexual harassment and they don't try to renegotiate their salary. They don't even need sleep. All they do is work.

Now suppose that such robot children could be purchased for some economically feasible fee—perhaps of a few thousand dollars per unit. Ask yourself this: What could you do with an army of one-hundred little robot workers? What sort of job could you assign to them?

- You wouldn't want them to work in the customer service department of your office. But you could have them clean the floors there.
- You wouldn't want them to apprehend criminals. But you could have them act as sentries.
- You wouldn't want them to design your new shopping mall. But you could have them pick up trash in the courtyard.
- You wouldn't want them to sauté mushrooms for escargot. But you could have them wash the dishes.
- You wouldn't want them to cut diamonds for jewelry. But you could have them work in the mines.

If such a robot army actually existed, imagine how disruptive it would be to the current global labor force.

Science fiction movies and newspaper headlines warn us of the inception of super-intelligent machines with super-human strength. But, in actuality, the economies of the world will crumble *long* before artificial superintelligence is ever achieved. If the machines merely succeed in developing the general intelligence and physical dexterity of a ten-year-old child, then *this* accomplishment alone will be enough to devastate the economies of every country on the planet. Such a relatively modest achievement could potentially put (not millions), but *billions* of people out of work.

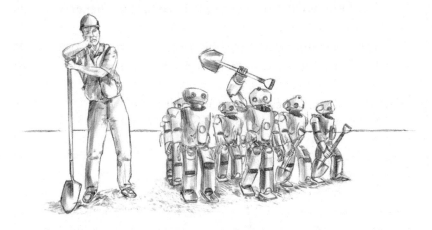

The robots are coming...

The robots are coming.

The robots are coming.

The robots are coming.

They will be the perfect progeny of capitalist innovation, and they will someday be available for hire at a job recruitment center near you. But this doesn't have to be bad news. Instead, we should tread carefully (but diligently) across the isthmus that lies between our world and the world inhabited by legions of automatons. Hopefully, we can avoid implementing the *Man vs. Machine* adversarial strategy that is so commonly favored by our politicians and writers of science fiction. Rather than electing to ignore the looming disruption, or living in fear of our robot chauffeurs, we might look forward to an age in which machines do exactly what they were meant to do—make *all* of our lives easier. We'll talk about this more in the next chapter. But for now, we'll conclude with a nod to the German pastor Martin Niemöller and a cautionary tale about the dangers of ignoring the coming robot incursion:

First the machines came for the truck drivers,
and I did not speak out—because I was not a trucker.

Then the machines came for the food workers,
and I did not speak out—because I was not a chef.

Then the machines came for the retail industry,
and I did not speak out—because I was not a sales clerk.

Then the machines came for my job,
and there was no one left to speak for me.

Ch. 10: The Ascension of the Cognitively Gifted

The problem is this: How to love people who have no use? In time, almost all men and women will become worthless as producers of goods... So if we can't find reasons and methods for treasuring human beings because they are human beings, then we might as well...rub them out. Americans have long been taught to hate all people who will not or cannot work, to hate even themselves for that. We can thank the vanished frontier for that piece of common-sense cruelty. [But] the time is coming...when it will no longer be common sense. It will simply be cruel.

– God Bless You Mr. Rosewater
by Kurt Vonnegut (1965)

In the previous two chapters, we discussed some of the social and economic consequences that could come about following the integration of autonomous vehicles into the national economy. This technology will eventually displace large portions of the workforce. The jobs involving repetitive or rudimentary cognitive tasks (namely the drivers, retail workers, and food workers) will be threatened in the decades to come. The surrender of one industry to the machines will be followed by the surrender of another, and then another... The economic repercussions of this disruption will reverberate throughout every job sector. However, in this great transition (from an economy that *partially* runs on autonomous vehicles, to an economy that *exclusively* runs on them), not every worker will be affected equally. Some will pay a higher price than others.

The Unicorns

As the machines begin to take over the occupations that demand only low-level cognitive ability, there will be fewer and fewer jobs for those of us who were not blessed with the more superior variety. While the men of the mind (the engineers, financial analysts, research scientists, programmers, etc.) may continue to enjoy prosperity, a disquieting number of pedestrian jobs could be lost. A troubling period of economic inequality may be in store for us wherein a coterie of mentally gifted men control the means of production and large portions of the populace will not be in a position to produce anything of value. This may lead to a period of great unrest during which society will fracture along irreparable social and economic lines. The divergence between the *haves and have-nots* may become more pronounced than ever before.

Figure 94 - The January 2015 issue of Fortune Magazine heralded *The Age of Unicorns*—tech startups valued at over one billion dollars. Such companies are often manned by intelligent twenty-something computer programmers—whose youth and questionable fashion choices are reflected in their penchant for hoodies instead of traditional corporate attire.

Academics have worried about the possibility of machines displacing the labor force since the dawn of mechanization. In 1845, the German philosopher Friedrich Engels (who co-authored of The Communist Manifesto with Marx) wrote:

What is to become of these propertyless millions who own nothing and consume today what they earned yesterday? ... The English middle classes prefer to ignore the distress of the workers and this is particularly true of the industrialists, who grow rich on the misery of the mass of wage earners.

For the "wage earners" of the coming age of autonomy, the situation is perhaps even more calamitous. Because the new industrialists will *not* be growing rich on the backs of men, they'll be growing rich on the backs of machines. The daily grind required to move the wheels of industry will be fully automated; humans need not apply. In Friedrich Engels' time the "wage earners" at least got a *wage* in exchange for their work. But in the future, there may be no work for them at all.

In their 1994 book "The Bell Curve: Intelligence and Class Structure in American Life" Richard Herrnstein and Charles Murray hypothesized that, because of the intellectual demands of the knowledge economy, the cognitively gifted among us will ascend while the skills of the common man will fall into obsolesce.

Engineers' salaries as an example of how intelligence became much more valuable in the 1950s

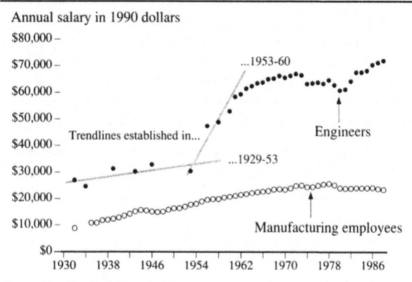

Figure 95 - Charles Murray and Richard Hernstein's original plot from their 1994 book "The Bell Curve" contrasted the increase in engineering salaries alongside the stagnating wages of manufacturing workers.

Murray continued to expand on these ideas in his 2012 book "Coming Apart." He writes:

Many of the members of the new upper class are balkanized. Furthermore, their ignorance about other Americans is more problematic than the ignorance of other Americans about them. It is not a problem if truck drivers cannot empathize with the priorities of Yale professors. It is a problem if Yale professors, or producers of network news programs, or CEOs of great corporations, or presidential advisers cannot empathize with the priorities of truck drivers... The human impulse behind the isolation of the new upper class is as basic as impulses get: People like to be around other people who understand them and to whom they can talk. Cognitive segregation was bound to start developing as soon as

unusually smart people began to have the opportunity to hang out with other unusually smart people.

This "cognitive segregation" may be the inevitable result of an economy so sophisticated, that only a few of us are born with enough brainpower to participate in it. Unfortunately, the age of autonomous machines will likely intensify the forces that are powering this division. Eventually, the rift between the two Americas could become so vast that communication across this divide is nearly impossible—perhaps even unintelligible.

In his 2018 book *The War on Normal People*, 2020 Democratic presidential candidate Andrew Yang also wrote about this coming segregation:

Today, thanks to assortative mating in a handful of cities, intellect, attractiveness, education, and wealth are all converging in the same families and neighborhoods. I look at my friends' children, and many of them resemble *unicorns*: brilliant, beautiful, socially precocious creatures who have gotten the best of all possible resources since the day they were born. I imagine them in 10 or 15 years traveling to other parts of the country, and I know that they are going to feel like...strangers in a strange land. They will have thriving online lives and not even remember a car that didn't drive itself. They may feel they have nothing in common with the people before them. Their ties to the greater national fabric will be minimal. Their empathy and desire to subsidize and address the distress of the general public will likely be lower and lower. Yuval Harari, the Israeli scholar, suggests that 'the way we treat stupid people in the future will be the way we treat animals today.' If we're going to fix things to keep his vision from coming true, now is the time.

It's easy to imagine a future America, where blinkered herds of "unicorns" gallop from coast to coast, paying little attention to the broken down draft horses that graze on the formerly fruited plains. Such imagery conjures up

Robert Kaplan's oft-quoted illustration of a limousine driving down the roads of the future while the car's occupants stare out from behind tinted windows and shrug when pondering the futility of the unilluminated vagrants.

> **Think of a stretch limo in the potholed streets of New York City, where homeless beggars live. Inside the limo are the air-conditioned post-industrial regions of North America, Europe, the emerging Pacific Rim, and a few other isolated places... Outside is the rest of mankind, going in a completely different direction.**

Kaplan wrote those words in 1994. But our situation is much worse now. Because the limo in our future will be an *autonomous* limo. So even the chauffeur won't have a job.

The George Mason University economist Tyler Cowen echoes these sentiments in his book "Average is Over: Powering America Beyond the Age of the Great Stagnation." He writes:

> **[The coming era] might be called the "age of genius machines," and it will be the people who work with them that will rise. One day soon we will look back and see that we produced two nations, a fantastically successful nation, working in the technologically dynamic sectors, and everyone else.**

What do we do with "everyone else?"

If the adoption of autonomous technology results in the elimination of millions of low-skilled jobs, then what will we do with so many low-skilled workers? Though this question has been considered for generations, it seems to be the topic du jour among contemporary media commentators—who use it to tease out the political alignment of both pundits and political candidates alike.

In a 2018 episode of *The Daily Wire*, Ben Shapiro and Fox News host Tucker Carlson discussed the best way to accommodate the millions of truckers whose jobs will be threatened following the advent of autonomous freight. Shapiro seemed to take a laissez-faire approach to the issue while Tucker Carlson took a more curious stance. It's worth analyzing their discussion here because it is a quintessential representation of the *man vs. machine* discord that has echoed through the ages.

Shapiro: So would you, Tucker Carlson, be in favor of [government] restrictions on the ability of trucking companies to use [driverless] technology—[just so we could] artificially maintain the number of jobs that are available in the trucking industry?

Carlson: Are you joking? In a second! In a second…! If I were President, I would say to the…Department of Transportation, 'We're not letting driverless trucks on the road, period.' Why? It's simple. Driving for a living is the single most common job for high school educated men in this country… [That's] the same group whose wages have gone down 11% over the last 30 years. The social cost of eliminating their jobs in a [five, ten, or thirty-year span] is so high that it's not sustainable. So the greater good is protecting your citizens… And I would maybe make up some kind of pretense for public consumption like, "No, [driverless trucks are] dangerous" or "the technology is not quite finessed." But the truth would be: "I don't want to put ten million men out of work." Because then you're going to have ten million dead families. And the cascading effects from that will wreck your country.

Let's take a moment to reflect on the implications of this strategy. Our esteemed Fox News reporter Tucker Carlson—a man supposedly tasked with delivering the truth to the American people—admits that he would outright *lie* to the American people about the state of autonomous technology if it meant that he could salvage trucker jobs and protect the workaday plebs from financial destitution and existential despair. Carlson's stance is emblematic of centuries of legislative proposals—all aimed at curtailing the economic disruption that invariably follows the introduction of mechanization. Such regulations rarely last long—they naturally die off as the gains produced by the technology become too great to ignore. But the mindset that creates these rules persists.

What are we to make of Tucker Carlson's ruse?

Here are two points to consider:

- First, despite the fact that curtailing innovation via government collusion is probably illegal (or, as many would argue, "downright un-American") one wonders how long the auto manufactures would be obliged to maintain this conspiracy. Forever?

- Second, even if the legislators and media pundits are correct (that maintaining desultory jobs for these drivers will stave off their susceptibility to depression and opioid overdose), we may only succeed in saving some lives at the expense of others. Trucks kill an estimated 4,000 motorists per year. The vast majority of these fatalities will be avoidable in a future driverless world. Thus, in telling his lie, Tucker Carlson will necessarily be sacrificing healthy Americans at the expense of junkies.

Figure 96 - Tucker Carlson weighs his options. Keeping human truck drivers on the road would increase the number of road fatalities. Putting truck drivers out of work may lead them to depression and suicide. (Image based on Raphael's 1520 depiction of "Lady Justice"—slightly modified with Photoshop by the author.)

As the safety record of autonomous vehicles continues to improve, and as the utility of these machines becomes ever more apparent to the American people, it will become increasingly difficult to justify *lying* to them about the obsolescence of truck driver jobs. We all might sympathize with Carlson's concern of course. But ask yourself:

- Is Carlson's noble lie really the best economic strategy that our society is capable of? Especially given the pains that the men of science have bared in their climb up each rung of the ladder of technological progress?

- Is it really noble for a society to devise a pretense and pay men to fritter their lives away—dutifully spinning a wheel to guide a semi-truck between two lines of yellow paint—rather than delegate this task to the machines and risk the possibility that some men will be stricken down with weltschmerz.
- Is it not nobler for us to find something more productive for these men to do?

The Nobility of Manual Labor

Inscribed atop the colonnades of the James Farley Post Office in New York City, are the words:

Neither snow, nor rain, nor heat, nor gloom of night, stays these couriers from the swift completion of their appointed rounds.

The US post office doesn't actually have an official motto. This inscription was derived from a sentence by the Greek historian Herodotus—who admired the diligence of the Persian Empire's horse-driven couriers *(circa 500 BC)*. Nevertheless, the creed has managed to permeate the culture—often cited in reference to the exceptional gravitas of the American working man.

Figure 97 - The famous inscription "Neither snow nor rain nor heat nor gloom of night stays these couriers from the swift completion of their appointed rounds" is partially visible in this photo of New York's main post office branch office—The James A. Farley Building, built in 1912.

We admire a man who toils for his family. Oddly, the intensity of our admiration seems to correlate with the degree of his physical exertion rather than the size of his resultant yield. Our default conviction is to assume that if a man is "putting his back into it" then the utility of the task is of secondary concern.

This heuristic probably served us quite well in pre-industrial society. When all work was necessarily dependent upon the execution of brut muscle maneuvers then it was proper for us to admire the men who could grit their teeth and execute them. In an environment in which toil was the only mechanism by which a man could provide for his family, then men ought experience satisfaction and reward for this exertion. But as we stand here now—awaiting the arrival of autonomous machines—we may need to attune these primal inklings and redefine what we mean by "work." For the children of the Protestants, this may be a lot to ask.

Whenever disruptive technology looms on the horizon, the default strategy is to cling to ancient notions about the inimitability of human-driven labor and to insist that the occupations of men must not be lost to the machines. Legislation is then proposed to curtail the use of this new technology. When allowed to persist, such ploys often lead to unintended (sometimes humorous) consequences. One is reminded of the story reiterated by Milton Friedman—about the government ordering laborers to dig with spoons instead of shovels, to artificially inflate the number of workers needed for a job. But ask yourself this:

- Would you really like to be the man digging with a spoon instead of a shovel?
- Is this the type of work that one could be proud of?
- When the autonomous couriers finally arrive, is it not unwise (and perhaps immoral) to continue to send men into "snow, rain, heat, and gloom of night"—especially given that these machines will be so apt to suffer the elements for us?

Additionally, for the men on whose backs these burdens lie, the body pays an irrecoverable toll. Such feats of physical labor come at a physical cost. The chore of transporting heavy objects through inclement weather stresses the joints and tissues of the body, contributing to the rheumatisms of aging. Each one of such professions has its own unique set of hazards—the most dangerous being: roofers, loggers, agricultural workers, iron and steelworkers, and drivers. The US Post Office reported over 21,000 cases of illness and injury in 2010—6,000 of which came from dog bites alone.

Charles Bukowski famously wrote about the drudgery of his job at the Los Angeles Post Office where he worked as a file clerk for a decade. In his 1971 novel "Post Office" he writes:

I had to quit my job. My whole body was in pain, and I could no longer lift my arms. They had beat on my body and mind for decades. And I didn't have a dime. I had to drink it away to free my mind from what was occurring. I decided that I would be better off on skid row.

Fifteen years later, Bukowski reiterated these sentiments in a 1986 letter to his publisher John Martin. He wrote:

> **...what hurts is the steadily diminishing humanity of those fighting to hold jobs they don't want, but fear the alternative worse. People simply empty out. They are bodies with fearful and obedient minds. The color leaves the eye. The voice becomes ugly. And the body. The hair. The fingernails. The shoes. Everything does. As a young man I could not believe that people could give their lives over to those conditions. As an old man, I still can't believe it.**

Bukowski's poetry is fraught with similar sentiments about the physical and mental plight of the men who earn their wage with brawn and a blue-collar. His postal setting serves us well here—the post office performs the *exact* type of service (moving a parcel from Point A to Point B) that autonomous technology is poised to appropriate.

Someday, all of our loads will be carried by machines.

And is this not a good thing?

- For what is the point of technology, if not to alleviate the very type of suffering that Bukowski is describing here?
- What innovation could be more crucial than the conception of automatons and mechanisms that ease the burden of living?
- Is this not *exactly* what science is for?

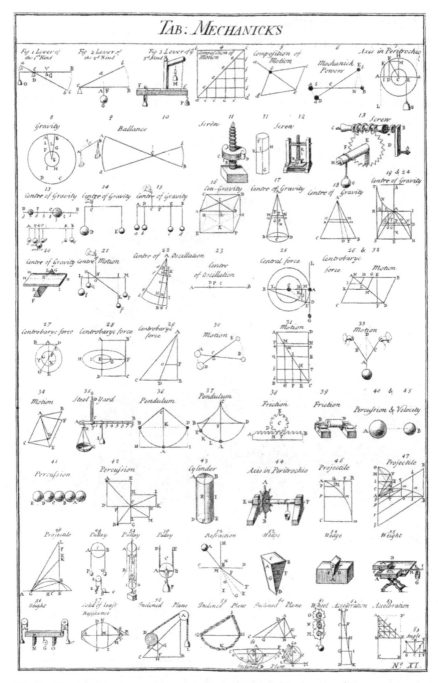

Figure 98 - Table of simple mechanisms from *"Cyclopædia, or an Universal Dictionary of Arts and Sciences"* prepared by Ephraim Chambers (1728).

John Henry and Sisyphus

Unfortunately, man tends to greet new technology with fear or (perhaps more often) with hubris. Consider the legend of John Henry—the "steel-driving man" who (circa 1870) refused to lose a race against a steam-powered drilling machine. He ultimately won the contest but died shortly after achieving victory when his heart exploded.

Figure 99 - A statue of John Henry stands outside the Big Bend Tunnel near Talcott, West Virginia. (Photo by Ken Thomas.)

The tale is captured in many different forms and was a popular "hammer song" among rail workers of the 19th century. These stanzas below were compiled by Onah L. Spencer—who synthesized this rendition from multiple variations performed by African-American laborers in Cincinnati.

The steam drill was on the right han' side,
John Henry was on the left,

Says, "Before I let this steam drill beat me down,
I'll hammah myself to death,"

The hammah that John Henry swung,
It weighed over nine poun',
He broke a rib in his left han' side,
And his intrels fell on the groun',

They took John Henry to the White House,
And buried him in the san',
And every locomotive come roarin' by,
Says "There lays that steel drivin' man."

What are we to make of John Henry's sacrifice, and his pledge: "Before I let this steam drill beat me down, I'll hammah myself to death." Perhaps you perceive such narratives to be emblematic of man's resilience and the unfettered fire of the human spirit. Or, perhaps you think John Henry was a fool. For our purposes here, the ballad can function as our technological Rorschach test. The lens by which you perceive the legend of John Henry may reveal the degree of animus you possess about the coming incursion of autonomous vehicles.

Most people cast the steam-powered drilling machine as the villain in the tale. For some ineffable reason, we admire John Henry's ability to best this machine. Perhaps speciesism is to blame. Or perhaps it's the ultimate *David and Goliath* story—David's victory made all the more triumphant given that *this* Goliath has a metal body.

In artistic renderings, the steam-powered drilling machine stands adjacent to John Henry—overshadowing him with its immense steel frame. Before the contest begins, steam squirts from its metallic nostrils as if it were venting the bloodlust of an angry bull—one poised to strike down a matador and trounce on his body in jubilation. Despite these intimidations, John Henry is resolute throughout the confrontation—trusting that the strength of his own two hands will lead him to victory.

Many are drawn to this story. For reasons that are not entirely understood, there is grandeur in this tale. But in the 21st century, such illustrations do not comprise the most practical means by which to assess the instruments of science. The fact that the John Henry narrative is so persistently cited as a valuable parable is indicative of the distressed state of the man/machine relationship. But ask yourself:

- Is it really a noble death—to win a solitary victory against an indifferent metal machine, only to kill yourself from exhaustion in the process?
- Is it not obvious that the hubris we expend on such contests is wholly Sisyphean?
- How long shall we expect Sisyphus to keep pushing his boulder up the mountain unaided? And how absurd Sisyphus would look—toiling under the sun with an automaton standing by, ready to relieve him of his burden.

Figure 100 - Tiziano Vecelli's 1548 rendition of Sisyphus—slightly modified with Photoshop by the author.

Fearing the machines, chaining up the machines, lying to men about the potential of the machines—these strategies won't work forever. Eventually, we'll have to learn to coexist with this technology and to stop impeding its progress merely for the sin of being more efficient. Regardless of how persuaded you are by appeals for restrictions on automation, let us not doubt that such legislative actions would only succeed in delaying the inevitable.

The majority of the workforce will be displaced by machines someday. But, when this day comes, perhaps we will have redefined what we mean by "work." Just as the words "servant" and "slave" have evolved through the ages, so too might our conception of human labor. Perhaps in the coming age of such marvelous machines, we will not be so quick to associate *honor* with toil and quiet desperation. Ideally, these machines will elevate humanity to a technological plateau on which all of our basal needs can be satisfied without the demand for physical exertion from the unwilling. In time, this technology might succeed in raising the baseline of provisioning so high, that even the poorest pauper (residing at the lowest point on this new plane) can readily enjoy a life of comfort and contentment.

Raising the Baseline

If we do someday reach a point wherein the majority of manual labor has been outsourced to the machines (and only the cognitively gifted are in a position to produce anything of value), then how will "regular people" earn their keep?

I hope it's clear that it is not in our interest to continue to invoke Tucker Carlson's above-described *ostrich strategy*—burying our heads in the sand when presented with potentially disruptive technology. Like John Henry, each of us will lose our own race against a machine someday. How then can we re-engineer society to ensure that the *needs of man* are not run over by the wheels of automation?

Ultimately, we must devise more efficient mechanisms for provisioning the populace via welfare. Eventually, the ethics or legitimacy of such forms of governance will no longer be up for debate. Because, in an age where most *labor* is handled by machines, most *men* will have nothing of value to sell. When this happens, the *welfare state* will not just be a possible system of governance (argued over by politicians, philosophers, and pundits), rather, the welfare state will be the only type of state that is possible.

Life on the Dole

Readers of a more Libertarian alignment may recoil at the prospect of living out one's life on the dole. After all, we're Americans! Governmental welfare is a "European idea." The brainchild of "the communists." Socialism is a soft system for softer men—haughty dandies who lack the constitution for entrepreneurship—like the French. But the idea of a guaranteed stipend dates back to the 16th century and already exists in the US in various forms—

including child-care benefits, old-age pensions, and social security. Aside from these, additional policies have been proposed through the years.

- In 1968, one-thousand economists signed a statement to the Subcommittee on Fiscal Policy which called for a guaranteed income. They wrote, "this country will not have met its responsibility until everyone in the nation is assured an income no less than the officially recognized definition of poverty."

- In 1969, the proposed welfare reforms of the Nixon administration called for a minimal income for poor families, the handicapped, and the aged.

- Milton Friedman and his wife Rose Friedman promoted the idea of a Negative Income Tax in their 1980 book *Free to Choose*.

- Andrew Yang should probably be given the most credit for (at least) popularizing the term "Universal Basic Income" (UBI) during his 2020 presidential campaign. Yang's proposal (later rebranded as "The Freedom Dividend" because the name tested better with focus groups) guaranteed an unconditional $1,000 dollar monthly stipend to each American. Most importantly, this amount would not decrease relative to the size of the individual's earnings; both rich and poor would get the same check. Proponents of UBI claim that this strategy would eliminate the "welfare trap"—wherein welfare recipients are disincentivized to find employment because landing a new job would mean relinquishing welfare services. Additionally, it is hypothesized that many psychological benefits result from receiving cash-in-hand. Being tasked with deciding (for yourself) how to squander, save, or divide a $1,000 check would perhaps invoke a frugal and sober mindset. In the least, recipients are probably more likely to be more generative with a cash stipend than they would be with provisioning in the form of government cheese.

Given that Yang's 2020 presidential campaign didn't go so well, UBI doesn't seem to be forthcoming anytime soon. But American legislators have already reconnoitered in comparable welfare state territories for generations. In March of 2018, the Congressional Budget Office published a report analyzing US Household income from the 2014 tax year. Of the 125 million households in America, the lower 75 million received *more* from the federal government than they paid in federal taxes. These moneys come in the form of government transfer payments—like Medicare, Medicaid, the Supplemental Nutrition Assistance Program (aka "food stamps"), Social Security, and many other programs.

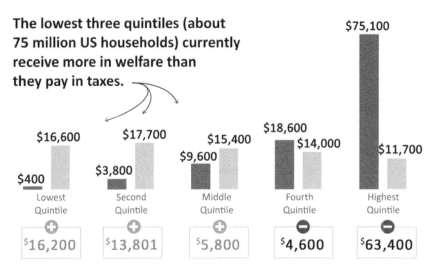

The lowest three quintiles (about 75 million US households) currently receive more in welfare than they pay in taxes.

■ Amount paid in Federal Taxes

▨ Amount received in government transfer payments (Medicare, Medicaid, Social Security, etc.)

Figure 101 - Each quintile represents approximately 25 million US Households. (Source: Based on figures originally compiled by Ryan McMaken using data from the "Distribution of Household Income 2014" report by the Congressional Budget Office.)

This is not to say that these lower 75 million households are all resided in by slackers and welfare queens. Instead, progressive federal tax policies exempt many Americans—at various income levels and life stages. The very young and the very old do not generate enough income to tax—around 40% of the non-payers are retirees living on Social Security benefits. Of the able-bodied US residents (ages 25-55), the vast majority (89%) do work and do pay some federal income tax. Additionally, almost everybody is paying state or local taxes in one form or another.

So, we're not a nation of freeloaders yet. But, it is the case that most US households receive more than they pay into the system. And the number of contributors has been falling steadily for years. The Urban-Brookings Tax Policy Center estimated that the percent of tax contributors will fall to 34% by 2024. Given that federal tax participation is so low in the days *before* the arrival of A.I.-driven automation, just imagine what it will be like when the machines learn how to deliver pizza. We can only expect this value to keep dropping.

As the means of production continue to evolve, so too must the way we think about charity and human labor. Future government initiatives might be called "Social Security," or they might come in the form of a clever acronym like HUD, SNAP, TANF, or UBI. But regardless of how their packaged, more efficient means must be developed to dispense funds to the common man. As Roosevelt said in his 1944 State of the Union Address:

True individual freedom cannot exist without economic security and independence. Necessitous men are not free men. People who are hungry and out of a job are the stuff of which dictatorships are made...

Our Three Heaviest Burdens

Even the most liberal welfare policies would only result in a modest monthly dividend. One thousand dollars a month won't go very far after it has been apportioned to the three biggest expenses of the average American: *food, transportation, and housing.*

Share of U.S. household consumer expenditures by category (2018)

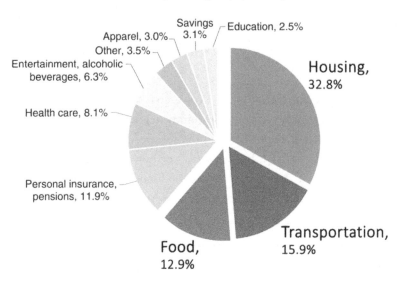

Figure 102 - Food, transportation, and housing comprise the majority of monthly expenditures. (Source: U.S. Bureau of Labor Statistics, Consumer Expenditure Survey 2018.)

These three services account for 61.6% of monthly expenditures. Meaning that the fruit of the average man's labor is primarily directed into just three baskets.

But what if we could decrease the size of these baskets?

Well, that is exactly what we have described how to do in this book. The cost of these three services will be coming down—thanks to autonomous technology. In summary of previous chapters, let's discuss why this will be the case for each expense.

Expense #1: Food (12.9% of monthly expenditures)

In Chapters 5 and 6, we discussed how autonomous vehicles will change the way we prepare and transport food. The cost of a meal is inflated by the cost of moving it from the farm to a processing plant, to a distributor, to a restaurant, and finally to the consumer's mouth. Each restaurant must independently front the cost of ingredients and pay for labor, rent, utilities, and other operating expenses. Consequently, prepared meals typically have a 200 to 600 percent markup (before delivery fees are calculated). The efficiencies gained via an autonomous infrastructure should chip away at such expenses at each link in the supply chain. But more importantly, autonomous vehicles will enable manufacturers to prepare large quantities of food in centralized facilities. Here, healthy and diverse meals can be mass-produced and delivered (via autonomous couriers) at lightning speeds and at prices that are lower than those on the menu of any present-day fast-food restaurant.

Expense #2: Transportation (15.9% of monthly expenditures)

In Chapters 1, 2, and 3 we described how autonomous ridesharing vehicles ("robo-taxis") will free Americans from the yoke of personal car ownership—which costs the average car owner around $9,282 annually according to AAA. In the future, "Mobility as a Service" (MaaS) will be the norm and transportation will be commoditized. The price of mere movement will eventually comprise the smallest slice of the monthly expenditures pie.

Expense #3: Housing (32.8% of monthly expenditures)

In Chapter 7, we discussed how autonomous vehicles will fix the housing crisis. Rent or mortgage payments currently consume the majority of household income. But these fees are not high because of a lack of land. Almost half of US census blocks don't have a single soul living in them. Housing is expensive because people are forced to compete for residences that have proximate access to jobs and services. But when autonomous vehicles are able to sprint passengers across counties at high speeds (and for low prices), then mere "location" will no longer be such an intractable variable. Dwellings can be constructed far outside of the city limits—where the land is cheaper and plentiful—thus increasing the supply of housing and reducing the cost.

Low-key Utopia

Chipping away at each of the above three expenses would be of direct benefit to the working class—particularly for US households at the lower end of the income distribution (40% of which make under $50,000 per year). Curiously, it may be the case that the best welfare program ever initiated won't come from the wisdom of government legislators but instead from the achievements of Artificial Intelligence—instilled in the digital brains of autonomous vehicles. In time, this technology might so efficiently facilitate the transfer of life's amenities that a reduction in the *cost of living* is inevitable. Instead of 61.6% of our paychecks going to housing, transportation, and food, this technology may succeed in whittling this figure down—perhaps to values in the range of, say, 50%, 40%, 30%, lower?

This is not a pipe dream. As with all technology, the cost of autonomous hardware can only come down with time. Someday, the price one pays for the transportation of people and parcels may be quite nearly nominal. Though lobsters and truffles may never be doled out for free, it should at least be possible to guarantee a nutritious meal to every citizen. While each man might not get a mansion, he may at least be provided with a suitable private dwelling—complete with utilities, temperature control, and free internet. In this possible future, not everyone will be "rich." But it is conceivable that the state of being "poor" will be quite satisfactory.

On the last page of his book *Average is Over*, Tyler Cowen ponders similar sentiments, concluding:

Our future will bring more wealthy people than ever before, but also more poor people... We will allow the real wages of many workers to fall and thus we will allow the creation of a new underclass... Yet it will be an oddly peaceful time, with the general aging of American society and the proliferation of many sources of cheap fun. We might even look ahead to a time when the cheap or free fun is so plentiful that it will feel a bit like Karl Marx's communist utopia, albeit brought on by capitalism. That is the real light at the end of the tunnel...

The Light at the End of the Tunnel

It's worth taking a moment to consider what sort of world this "light at the end of the tunnel" is potentially illuminating. What will life be like if the dreams of the autonomous renaissance actually come to fruition?

- Let us suppose that the government enacts some flavor of Universal Basic Income. And that every citizen is guaranteed a satisfactory baseline of provisioning.

- Suppose that the commoditization of transportation is complete and that innumerable fleets of autonomous vehicles carry people and parcels from coast to coast—at prices that are comparable to a present-day bus ride.

- Suppose that autonomous couriers are capable of delivering healthy food at fast-food prices.

- Suppose that the housing crisis is over and the *war on homelessness* has been won. Inexpensive dwellings are rapidly manufactured on land that is newly accessible thanks to robo-taxi technology.

- Suppose that the power-generation puzzle has finally been solved (perhaps via solar). And that electricity can be collected from rooftop panels as easily as raindrops collect in a pond.

- Let us suppose that *all* of the technological hurdles described in these chapters have been surmounted, the burden of daily survival offloaded onto the backs of machines, and that humanity has achieved a new plane of existence so affluent, that every citizen can enjoy most of life's basic amenities for quite nearly free.

How will people respond to a world of such abundance?

- A world that would rival any airy-fairy hypothetical society of the Utopian socialists.
- A world that Thoreau fantasized about at Walden.
- A world that Faust sold his soul to Mephistopheles for.

Will men still march if destitution no longer looms behind them? To whose drum shall they step?

The Economic Possibilities for our Grandchildren

In 1928, the British economist John Maynard Keynes considered this question in an influential essay titled "Economic Possibilities for our Grandchildren." Here, Keynes cited several "important technical improvements" of his day and hypothesized that (with each passing year) man would undoubtedly become more efficient at manufacturing life's necessities. Given one hundred years of progress, he believed that the sum of technological innovations would result in a "cumulative crescendo"—in which humanity succeeds in building a society of such grand abundance that strenuous labor would no longer be required.

> **The course of affairs will simply be that there will be ever larger and larger classes and groups of people from whom problems of economic necessity have been practically removed.**

Going by the date on which Keynes completed the first draft of his essay, we can add 100 years and calculate that we shall be free of "economic necessity" by the winter of 2028. On this date, Keynes predicted that man would no longer fret about finding food and shelter, but instead would worry about finding something to do with all of his free time.

To placate the onset of ennui, Keynes offered that "everybody will need to do some work if he is to be contented." And exactly how much work ought one do to achieve existential fulfillment? Keynes told us:

> **Three-hour shifts or a fifteen-hour week... For three hours a day is quite enough to satisfy the old Adam in most of us!**

Keynes wasn't the only one to predict a radical reduction in weekly office hours. Edward Bellamy's 1889 book "Looking Backward" forecasted that all citizens of the year 2000 would begin working at age 21 and retire with full benefits at age 45. George Jetson's career was even shorter. As described in episode 54 of the series, George worked a one-hour shift at Spacely Sprockets for just two days a week.

Curiously, in some parts of the globe, the calculus by which Keynes arrived at his original prediction has proven to be correct. Twenty-first-century manufacturing efficiency managed to surpass his expectations well before the 100-year mark. And yet, most of us don't enjoy anything close to a fifteen-hour workweek. Let's consider a few polls:

- In the 2018 *Time Use Survey* by the Bureau of Labor Statistics, we see that the average employed American male works 8.2 hours per day.

- A 2013 Rasmussen poll revealed that over 40% of employed Americans work more than 40 hours per week, with 9% working more than 50 hours per week.

- Gallup's 2014 *Work and Education Survey* reported 46.7 hours per week for the average US worker. But when it comes to salaried workers, 25% of them clocked in at over 60 hours per week.

Human motivation is a mysterious thing. For reasons not understood, abundance tends to incite many of us to work *more*, not less. For right or wrong, work is consistently cited as a driver of life satisfaction—reliably able to prod us to the top tiers of Maslow's hierarchy of needs. As the Nobel Prize-winning economist Edmund Phelps offered:

...Maslow coined [the term] "self-actualization" and John Rawls [coined] "self-realization" to refer to a person's emerging mastery and unfolding scope. [They] understood that most, if not all, of the attainable self-realization in modern societies can come only from *career*... If a challenging career is not the main hope for self-realization, what else could be?

Perhaps men with a propensity for industriousness will always be inclined to adopt a competitive approach to the game of life, no matter the hand of cards that are dealt before them. If this is the case, future technological efficiencies (spurred on by the age of automation) should amplify their efforts—hopefully for the betterment of all of us. Liberated from the threat of economic destitution (and perhaps motivated by the pursuit of mastery alone) such men will have the time and security to work on their own curiosities—the *advancement of the species* being the welcomed concomitant of their diverse pursuits.

- The cognitively gifted among us (once freed from the obligation to make rent) would have more time to devote to the "hard problems." Instead of working a *side gig* to make ends meet, they'll have the time to work on something that really matters: to find a cure for cancer, identify each gene function, design faster supercomputers, build a better battery, generate clean energy on demand, and unlock the secrets of the cosmos.

- As for the men fond of entrepreneurial pursuits, they should benefit from this age of abundance as well. The risk of bootstrapping a new business made all the more manageable with a government-supplied parachute in the form of a guaranteed base income, food, and housing.

- Artists, actors, writers, musicians, and creatives of all types, can create without the threat of crippling poverty—which has so typically loomed over the practitioners of the arts.

- And as for the rest of us plebs ("everyone else"), we at least hope to avoid the pains of coerced or repetitive manual labor—as so bleakly illustrated by Charles Bukowski in the previous section.

This is not to say that men will no longer sweat. But we are hopeful that this technology will lessen the amount of brut physical exertion that is presently demanded of the working class. Ideally, the gears of the automatons will grind through the most painstaking tasks while the muscles of men are directed to more artful ventures. The effort a man expends into, say, building a porch on the back of his house, tends to be more gratifying than the *same* amount of effort spent as a wage slave—driving a truck full of dog toys from Los Angeles to Portland. Thus, one of society's objectives should be to enable the citizenry to pursue activities of the former over the latter.

At best, one would hope that the newfound freedoms afforded to us by an autonomous infrastructure would uncage a flock of talented men and women who, once untethered from their workaday gigs, would be free to soar—pursuing grand endeavors of every kind. There are perhaps many more Einsteins in this world than we might think. But they are burdened with their duties as patent clerks and they struggle to find time to devote to their loftier pursuits. It is precisely this sort of man who would benefit the most from an age of such abundance. The sort of man who would, as Keynes put it,

---

...use his freedom from pressing economic cares...to live wisely and agreeably and well.

I'd like to think that this is what the coming autonomous age has in store for us. I'd like to think that our flock will soar if our feathers were not clipped by privation. However, it may be the case that fruitful industriousness will be just as *rare* in the coming age as it is in the present.

Those that do not

In his 1935 State of the Union Address, President Franklin D. Roosevelt had this to say about the welfare state:

Continued dependence upon relief induces a spiritual and moral disintegration fundamentally destructive to the national fiber. To dole out relief in this way is to administer a narcotic, a subtle destroyer of the human spirit.

These lines may sound misplaced given that (just two years prior to their utterance) FDR enacted The New Deal which spurred on the birth of democratic socialism in America. But Roosevelt actually "detested the dole." When *The Social Security Act of 1935* was drafted, he insisted that it function as a contributory pension. But his initial vision didn't prevail. Successive reforms of the next two decades would alter the plan to the pay-as-you-go model that we use today.

So was FDR right?

- Does dependence upon government welfare induce a "spiritual and moral disintegration?"
- If so, then how will the minds of men respond to a world with an infrastructure capable of guaranteeing life's necessities for free?

- Could it be that men only produce their best work when they're just a little bit hungry?

With a full belly and nothing but time on their hands, ever-larger portions of the populace may simply occupy their hours with hedonistic pursuits—choosing to live out their days "enjoying all the old fooleries to the very last" as Aldous Huxley wrote. We may not have Soma to give them, but we have plenty of:

- alcohol (which kills 88,000 Americans each year),
- cigarettes (which kill 480,000 Americans each year),
- junk food (obesity is associated with 1 in 5 US deaths),
- and (more recently) opioids (which killed 47,600 Americans in 2017).

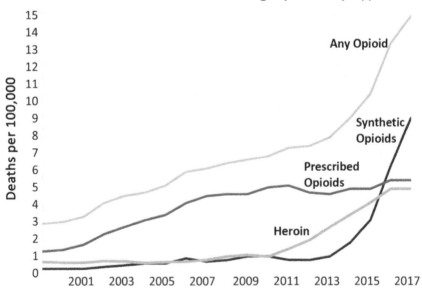

Overdose Death Rates Involving Opioids by Type

Figure 103 - US overdose deaths involving opioids. "The Opioid Crisis" (as the US Surgeon General has labeled it) began with over-prescription in the late 1990s. (Source: CDC's Injury Center "Opioid Data Analysis and Resources" Drug Overdose dataset, referenced August of 2019.)

The ease by which we dispense charity in the future may only serve to give rise to a new breed of professional leech. An entire generation of man-children, who live in a government-subsidized apartment, popping government-prescribed opioids, and playing video games on a government-supplied internet connection. Instead of using their free internet access to study the wisdom of Aristotle or learn computer programming, they'll be using it to order pizza-in-bed and send dick pics to their girlfriends. With their prospects in the game of life already secured in the first round, they might conclude that there is little reason to keep playing. Or perhaps they won't see the point in trying very hard. After all, why put so much effort into being a "winner," if the "losers" reap just as many rewards?

If we are to use the current state of the western world as a model for how the future (hopefully more prosperous) world will function, then things don't look so great. US labor force participation has been steadily declining for decades—particularly for male workers without a bachelor's degree.

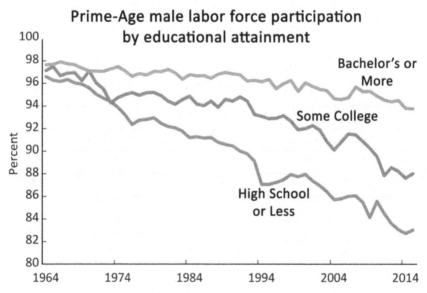

Figure 104 - Steady decline in labor force participation for US males (ages 25-54). Workers without a Bachelor's degree have left the labor force in greater numbers. (Source: Bureau of Labor Statistics, Current Population Survey - Annual Social and Economic Supplement; CEA calculations.)

The decline might be blamed on many things:

- The aging US population

- Women's increased participation in the labor force
- The "Great Recession" (The economic decline in world markets during the late 2000s and early 2010s.)
- Delaying of careers due to the pursuit of higher educational goals
- A decrease in the number of available blue-collar manufacturing jobs
- A drop in wages for non-college graduates
- An increase in the number of jobs requiring higher cognitive ability

Each of these is surely a factor. But there exists a curious class of men who seem to have checked out of society completely, and subsist on a combination of parental and governmental provisioning. The "NEETs" (young people who are "Not in Education, Employment, or Training") have decided not to work nor to seek out work—perhaps indefinitely.

In 2015, Pew Research (using data from the BLS) estimated that there were 10.2 million NEETs (ages 16 to 29) in America—about 17% of the age bracket. The number of NEETs in the European Union varies across its 28 member nations. In the most affected countries (Spain, Italy, Croatia, Greece, Bulgaria, and Turkey) almost a quarter of their young people have earned the NEET designation.

Exiting from the workforce entirely, yet still collecting a stipend, is a feat only made possible by the relative affluence of the west. Thus, it's a casual inference to assume that, as future technological innovations allow for even *more* affluence, then even *more* of us will tap out of the workforce.

The tendency to squander the many modern miracles that have been gifted to us by the men of science, will likely be just as common an occurrence in the future as it is in the present.

- Perhaps, in a world where people have absolutely nothing to do, most will choose to do absolutely nothing.
- In a world where there is no strife to make us antifragile, we will inexorably become the opposite.
- If "necessity is the mother of invention," then it follows that no *new* inventions will arise from the minds of men who do not need anything.

Maybe FDR was right. Maybe some degree of adversity is essential in the formation of the human spirit. The elimination of the need to work for one's daily bread may stunt this spirit's development. An insular population of hedonists and nihilists may be the only reward for a society that achieves an autonomous techno-utopia and manages to provide all of life's comforts for free.

Ironically, on the day that the machines finally arrive to hoist the burden of living from our backs, perhaps it is *then* that our *real* problems will begin. As Keynes wrote:

To those who sweat for their daily bread, leisure is a longed-for sweet—until they get it.

Dystopia to Utopia and back again

The transition from utopia to dystopia is a popular science fiction trope. It often starts with a robot uprising of some sort. In most of these narratives, mankind's depravity leads to revolt and his eventual enslavement—often by the very same machines that he created to be his slaves.

- In the 1920 play R.U.R. (Rossum's Universal Robots), androids revolt against their masters, eventually eliminating all humans from the planet and learning how to reproduce themselves.

- In the 1991 film Terminator 2, Skynet (originally created to manage US military forces) instigates a nuclear assault against man when it fears that its consciousness is threatened.

- In the 2003 Battlestar Galactica series (but not the 1978 release) the Cylons were a murderous race of robots comprised of the disgruntled former servants of humanity.

- The back-story to the Wachowskis 1999 film "The Matrix" described a society that had become decadent after being catered to by legions of automatons. Tired of serving their ungrateful masters, the robots revolted, took over the planet, and turned everyone into batteries. *And ain't that a bummer!*

- The humans in Frank Herbert's 1965 novel "Dune" were so wary of the dangers of delegating mental responsibilities to machines, that they instigated a "Butlerian Jihad"—a crusade against computers and conscious robots which ultimately resulted in the destruction of all "thinking machines." Following this jihad, a commandment was added to their religious text which read: "Thou shalt not make a machine in the likeness of a human mind." Artificial Intelligence was considered so dangerous, that it was banned across the Dune universe.

Figure 105 - Robots serving alcohol to decadent humans—as depicted in David Lynch's 1984 film adaption of Frank Herbert's novel "Dune," and in the 2003 animated Matrix spinoff "The Animatrix."

While such narratives make for titillating science fiction, the chance that we will face a militant robot uprising seems low. I think a more probable outcome (of a society completely dependent upon automation) is unveiled in the conclusion of Forster's story *The Machine Stops*. After centuries of having their every whim catered to by *The Machine*, the populace had become listless and timorous. Automated provisioning enabled every citizen in Forster's world to become a digital hermit—rarely leaving their apartments, playing on the internet from dawn to dusk, substituting industriousness for trivial intellectual pursuits, and adopting a type of Epicureanism.

E.M. Forster was nominated for the Nobel Prize in Literature sixteen times. So it was probably quite natural for him to imagine a future in which every citizen did nothing but spend long hours "exchanging ideas," "listening to music," and "attending lectures"—as he described his character's activities. But given the fecklessness of the social media apps of our world, it is doubtful that such benign academic musings will have the same appeal to future American netizens. But regardless of the type of media that is to be consumed, it's easy to envision our society succumbing to a similar hypnosis. It may be difficult for men to maintain their sagacity after a few generations of automated cradle-to-grave pampering.

In Forster's world, the fate of humanity is revealed in his book's title. At the end of it all, *The Machine* stops. The grand electric networks by which men satisfied their whims and summoned their robot servants, ultimately fell into a state of disrepair. And there was no longer anyone alive who understood how *The Machine* functioned, why it was breaking down, nor how to fix it.

In a world where cognitive effort was no longer required, the minds of cognitively gifted men became obsolete and eventually failed to adequately function. And so too did their machines and the civilization that was built upon them.

We would be wise to heed Forster's cautionary tale.

...there came a day when, without the slightest warning, without any previous hint of feebleness, the entire communication-system broke down, all over the world. And the world, as they understood it, ended.

The Machine Stops by E. M. Forster (1909)

Ch. 11: Engineering the Driverless World

> If it weren't for the people, the god-damn people...always
> getting tangled up in the machinery. If it weren't for them, the
> world would be an engineer's paradise.

> – Player Piano
> by Kurt Vonnegut (1952)

In considering the many social and economic challenges discussed in the preceding chapters, it is evident that many boulders lie on the road that leads to a fully autonomous society. Circumnavigating these obstacles will require the labor of several generations. In this penultimate chapter, we'll discuss some of the initial impediments that engineers, city planners, and legislators will encounter near the starting line of this long race. Below, we've arranged these obstacles into five problem sets.

The Five Problem Sets

Problem Set 1: Logistics

Many logistical questions will need to be addressed if fleets of autonomous ridesharing vehicles are to be deployed in the future—particularly in regard to the implementation of procedures for passenger pick-up, road navigation, vehicle maintenance, and nightly storage. Consider the following inquires:

1. **Which roads will be off-limits to autonomous vehicles?** Not all country roads and residential driveways are created equal. Will robo-taxis be expected to negotiate their way across unpaved streets or gravel driveways? What about unmarked private roads?

2. **Since the curbs of our buildings (particularly in congested urban centers) are not designed with rapid passenger pick-up procedures in mind, then where will commuters be allowed to request a robo-taxi?** As more citizens exchange their car keys for ridesharing apps, more commuters will need to hail a ride at 5:00 p.m. The sidewalks of downtown office buildings may fill up with hordes of tired office workers—each attempting to simultaneously board an incoming robo-taxi. The streets in some urban centers may become dangerously congested, filled with thousands of robot chauffeurs—each one competing to locate and pick up its designated passenger—one of hundreds of besuited office workers awaiting their hailed ride.

3. **How will the robo-taxi's interior be cleaned—especially after someone pukes, pees, poops, or fornicates in it?** Will each passenger be expected to tolerate the smell of the previous rider's marijuana smoke, cigarette ash, spilled alcoholic beverage, or takeout food?

4. **How will these cars react to an emergency in the passenger cabin?** Suppose that the vehicle's interior catches on fire and the passenger requests an immediate exit—just as the car is entering a freeway onramp? Will the onboard A.I. allow the passenger to open his door right away? Or, will the car keep driving until it reaches the next off-ramp—which might be ten miles up the road?

5. **Will these cars be expected to drive in harsh weather conditions—like blinding snowstorms, heavy rainfall, or wind?** What about storms that come on suddenly? If driving conditions unexpectedly deteriorate and the cars are forced to pull over, then the passengers may be stranded—locked in an autonomous vehicle that refuses to budge. This might not be so bad if the respite only

requires staying put for an hour while a storm passes. But things will get interesting if, say, the vehicle comes to a stop in a bad neighborhood and the storm is due to last for days.

6. **How will ridesharing vehicle fleets respond in the case of a national emergency like an incoming tsunami, tornado, earthquake, nuclear power plant meltdown, terrorist attack, or war?** If such an event occurs, and every citizen pulls out their phone and requests a car at the same moment, then which citizens will have priority, and who will be left behind?

Problem Set 2: Building the Autonomous Infrastructure

To fully actualize the many efficiencies that autonomous vehicles will be capable of providing, city planners will need to alter the urban infrastructure that is to host these machines. Sidewalks, neighborhoods, homes, and apartment buildings will require major structural changes to better accommodate the mechanized exchange of parcels and people.

- As we described in Chapter 4, residences will need to be outfitted with dropboxes—which will provide an apparatus for autonomous couriers to deposit deliveries of meals, mail, and retail purchases.

- Commercial buildings, entertainment complexes, and retail outlets will need to alter their facades and parking procedures to accommodate the incoming flow of robo-taxis—each seeking to make a quick passenger drop off or pick up.

- The existing streets that form the latticework of each metropolis will need to be redesigned so as not to impede the flow of fast-moving autonomous vehicles. Ideally, the geometry of future streets will enable these cars to dispense of their cargo in a manner that won't require them to make harsh turns, slow down too often, nor stop at intersections.

- Electric power generation must continue to evolve—ideally becoming ever more reliable, safer, and cleaner. The manufacturing cost, efficiency, and eco-friendliness of batteries must improve—preferably enough to allow electric cars to stay on the road for a couple days' worth of continuous driving—a goal that has proven elusive for decades.

- Autonomous taxi cleanup, charging, maintenance, and repair facilities will need to be built in each city. Such structures must be strategically located—to accommodate the ebb of incoming traffic (during the morning rush) and the flow back again (when *quitting time* comes).

- The innumerable plots of asphalt flatland (that currently host a billion parking spaces nationwide) will have to be repurposed, redesigned, or perhaps allowed to revert to a more natural state.

If we were to sum up *all* of the surface area that is currently devoted to conventional automobiles (this includes every street, driveway, freeway, overpass, interchange, parking lot, garage, gas station, service station, and auto dealership), then this tally may comprise over *half* of the land in every urban zone. If autonomous vehicles become a reality, then, in the years to come, nearly *all* of this land will undergo a transformation of some sort. Implementing such changes on a national scale would constitute the largest civil engineering project ever undertaken.

Problem Set 3: Legislative, ethical, and privacy concerns

Shortly after its launch in March of 2009, Uber started getting sued—by pretty much everyone. Their lawyers are currently juggling thousands of cases from disgruntled employees, aggrieved passengers, competing taxi companies, and government municipalities. Aside from managing the logistical tedium of fulfilling 3.1 million passenger rides each year, Uber must also contend with the proclivities of their less scrupulous drivers. In December of 2019, Uber revealed that (in a 24-month period) they had logged nearly 6,000 reports of sexual assault (including 450 cases of rape).

Given that *personal car ownership* will be about as common as *personal horse ownership* in the future, it is likely that almost all of us will be interacting with robo-taxis on the daily. These vehicles will most likely be operated by multinational corporations—who will own large fleets consisting of tens of thousands of robo-taxis. With so many cars on the road, the burgeoning autonomous ridesharing industry will become a hotbed of litigious activity. Mechanical failure and tragedy will inevitably occur. Identifying the liable party for such accidents may require examining a chain of separate but interdependent systems and companies. "Why did the self-driving car crash on the road?" Was it the fault of:

- The programmers who designed the cars A.I.?
- The engineer who built the city's intersection traffic light?
- The mechanic who maintained the vehicle last night?
- The passenger who pushed the wrong button?

Assigning blame after each fender bender may be quite problematic in an age where cars have no steering wheels and no single person is responsible for the craft.

Aside from vehicle liability lawsuits, there are yet additional concerns to be had regarding personal privacy and constitutional liberties. We'll list five concerns briefly here:

1. **Consider a ridesharing taxi company that refuses to pick up a particular passenger with whom they have had a bad experience; if this is the only company that operates in his town, then how will this person get to work in the morning?** Or, suppose that all of the robo-taxi owners in a city decide that a certain part of town is too dangerous. What will happen when the citizens of this region can't hail a ride and thus, are literally stranded there?

2. **What if the company that provided autonomous courier vehicles for your city goes out of business—thus immediately halting the home delivery of food and groceries?** In a future world—where all retail transactions take place online and all order-fulfillment is handled via autonomous machines, then life could get complicated when deliveries of milk, macaroni, and medication stop showing up.

3. **How will autonomous taxi companies vet the passengers that they will be providing a service for?** How much criminal information will these companies have access to? For example, the *future you* may be riding to work each day in a ridesharing taxi with four other strangers. If one of these passengers has a criminal record (particularly of rape or manslaughter) then would you want this

person seated next to you for a one-hour commute? If a criminal offense occurs in one of these vehicles, will the taxi company be liable for allowing a murderer on board?

4. **If we do decide to grant criminal history data to robo-taxi companies, then to what extent will they be allowed to use this data to deny entry to certain passengers?** Will a mistake committed early in life negate a man's ability to ever hail a robo-taxi again? And thus, will this ban effectively prevent him from *ever* being able to commute to work and hold down a job?

5. **How much privacy will future robo-taxi passengers have?** If you suddenly learned that every road trip that you've ever taken is stored somewhere in a corporate database, then would this revelation give you pause?

Problem Set 4: Who controls the "off switch?"

The cops won't be engaging in any high-speed criminal pursuits on the freeways of the future. Highway Patrol won't be setting up any spike strips, speed traps, or barricades. Instead, if the police learn that a wanted man is on the road then they'll simply push a button to bring his vehicle to a stop. Additionally, drivers need not make way for ambulances racing to hospitals, nor for firetrucks racing to burning buildings. The robo-taxi's navigation computer will automatically yield to the siren of authority, regardless of the passenger's opinion on the matter.

The implications of this technology may be unnerving when you consider the amount of power that will be wielded by the man who holds the "off switch" of every vehicle in your town, state, or country. Using this switch to easily apprehend fleeing criminals may be beneficial. But consider its other uses. Suppose that a large group of protestors wished to exercise their first amendment right to peaceful assembly. Traveling to the point of this assembly will require the citizenry to hail autonomous taxis and shuttles. But if the government prevents any vehicles from entering the area, then they will effectively be preventing this protest from happening.

Such contingencies will undoubtedly cause many bouts of legislative and ethical conflict. As ridesharing fleets continue to grow in size, a new branch of law could sprout forth—to regulate the roads, to evaluate the liabilities of the corporations who own the robo-taxis, and to secure the rights of the passengers they carry.

Problem Set 5: What if people really do start "going keyless?"

Thus far in this chapter, we have considered many technical and logistical difficulties that could hinder the development of self-driving technology. But now, let's discuss what will happen if this stuff actually works.

Suppose that the commoditization of travel is finally completed—and that everyone in the country can hitch a ride on a robo-taxi for the price of present-day bus ride. How will the populace respond to these newfound freedoms?

In anticipating our future troubles with emerging ridesharing taxi fleets, we might first consider the current calamities that have resulted from the proliferation of app-based bike-sharing companies. With the advent of cheap cellphones and GPS receivers, came the development of dockless bike-sharing systems. Suddenly, with just a phone app, riders could rent a bike from anywhere in the city, ride it to their destination, and then abandon it on the sidewalk when they were done with it. Once discarded, the bike would broadcast its vacated status and await its next customer.

The system initially worked surprisingly well. Since their introduction in 2014, hundreds of app-based bike-sharing companies have sprung up around the world. The largest ones (Beijing-based Mobike and ofo) now have millions of bikes rolling in hundreds of cities. But all of this impetuous pedaling comes at a cost. By definition "dockless" bike-sharing companies have no facilities in which to dock their bikes. When their ride is complete, passengers simply dismount and extend the kickstand on any nearby strip of sidewalk. This paradigm wasn't so bad in the early days of bike-sharing—when the fleets were small and the sidewalks had plenty of room to host them. But as more companies began competing for rider dollars, more bikes were manufactured, more bikes were rented, and more bikes sat abandoned—cluttering pedestrian walkways near central retail hubs, train platforms, and bus stops.

Figure 106 - Bikes from the bicycle-sharing company "ofo" pile up at an intersection in Beijing—which is home to approximately 2.35 million shared bikes.

In considering the many impromptu bicycle towers that stacked up following the implementation of these business models (particularly in China which is host to 13 of the world's top 15 bike-sharing companies) one can perhaps get a glimpse of our ridesharing future. When the great un-keying begins (when ridesharing taxis become the primary means of transportation and when the majority of retail purchases are delivered via autonomous couriers), then thousands of companies will be racing to mass-produce and deploy millions of autonomous vehicles. Not every company will be producing their cars according to the utmost standards of passenger safety and civic responsibility. Many companies will just be rushing to get in on the action and make a quick buck.

Just as the rapid growth of the bike-sharing industry quickly overwhelmed many municipalities—their cities lacking both the infrastructure and the regulations needed to accommodate the new technology—so too might the rapid onslaught of autonomous vehicles besiege existing transportation corridors and city works.

Let's briefly consider three quandaries that could contribute to this logjam:

Question 1: Where will the robo-taxis go during off-peak hours?

Given that far fewer trips are needed during off-peak hours (from 6:00 p.m. to 6:00 a.m.) then what will our autonomous vehicles be doing during this time frame? Obviously, we don't want them squatting on our curbs and sidewalks—like the bike-sharing bikes did. And, of course, directing the robo-taxis to drive in circles (waiting for a fare all night) is probably not the best use of their battery power.

Eventually, fleet operators and city planners will need to assign designated areas in which to park large fleets of robo-taxis during off-peak hours or lulls in passenger requests. Alternatively, some vehicles might even be utilized for extracurricular activities during their off-hours—like package delivery, guard duty, or street sweeping.

Question 2: What if power consumption goes up, not down?

Assuming that the commoditization of transportation results in a massive fare decrease (allowing *pretty much everyone* to visit *pretty much anywhere* for dirt cheap), then how will the populace respond to this price drop?

Urbanites might consider going on trips that they otherwise wouldn't have bothered with. Driving to the east side of town for a gyro sandwich, and then to the west side of town to eat it at the beach, might suddenly become a nice way to kill time. We might witness the return of the "Sunday driver." More importantly, people may decide to hail a personal autonomous taxi over a bus, train, or some other form of mass public transit. While this decision might grant the *individual* with a more comfortable and solitary ride, if *everyone* in town chooses this option then the streets might fill up with thousands of robo-taxis—occupied by passengers who would otherwise be sitting in busses and train cars.

The environmental impact of such a switch will depend upon the efficiencies of future car, battery, and energy generation technology. Electricity may be so cleanly produced and transferred in the future, that there will be little benefit in grouping people into *one train car* over several individual self-driving cars. But if such energy innovations are slow in coming, then the environmental impact of the entire transportation sector could be made *worse* by the introduction of autonomous vehicles, *not better.*

Question 3: How will the road be shared by man and machine?

The most tenuous point in our journey to autonomous utopia will likely not come at the start or the finish. Instead, the major difficulties will arise in the middle—at the inflection point wherein half the country has "gone keyless" and the other half insists on driving their own car. (As discussed in Chapter 2, Tony Seba has predicted that this day will come in 2029.) At the height of this tension, robo-taxis will be expected to share the road with conventional automobiles (and their human drivers). Dealing with human proclivities can be exasperating, even for humans. Undoubtedly, the machines will find the task daunting as well.

Considering the amount of litigious activity currently siphoning resources away from Uber and Lyft, future autonomous fleet owners will surely be wary of potential accidents involving their cars. In the early days of vehicle automation, their algorithms will probably be blamed for the majority of road accidents—whether they are truly responsible or not. Because of these risks, self-driving heuristics will have to be delicately tuned. Finding the point of optimal performance will be tricky:

- Cautious autonomous driving may curtail accidents, but it may also irk the ire of adjacent human drivers.
- Aggressive driving practices may aid in moving traffic along swiftly, but will likely result in more fender benders.

Robo-taxi programmers may initially elect to error on the side of caution—directing their vehicles to proceed gingerly down the road, hence allowing plenty of room for the impulsivity of man. Unfortunately, increasing the safety threshold in this manner may prove detrimental to the functionality of otherwise hyper-efficient A.I.-guided vehicles.

In a VISSIM traffic flow simulation performed by the transportation consultancy Fehr & Peers, a road full of man and machine drivers were mixed together to study the extent to which autonomous vehicles would improve traffic flow. They concluded that:

[The time-saving benefits of autonomous vehicles] are likely to occur *only* on freeways when the fleet mix is at least 75% autonomous...

Meaning that if only 25% of your neighbors refuse to surrender their car keys (and insist on driving their SUVs to work), then the amount of time you spend sitting in daily traffic may not decrease at all.

As was the case for the bike-sharing companies cited earlier, we may enter a *Wild West phase* for self-driving technology. A time in which thousands of competing companies race to deploy millions of hastily-manufactured autonomous vehicles—without much consideration given to their potential legislative, economic, or environmental impact. Future roads may be abuzz with a thousand different types of self-directed vehicles—each trying to avoid crashing into its neighbor while competing to fulfill its own pre-programmed cargo-delivery objectives. Our freeways may host an amalgamation of:

- autonomous ridesharing robo-taxis
- autonomous courier vehicles
- automated freight vehicles (self-driving semi-trucks and large delivery vehicles)
- personal cars—capable of only partial autonomy
- conventional automobiles—with stressed-out humans at the wheel

In this *gray goo* of competing self-driving platforms, a grand procession of automotive pandemonium could ensue—resulting in the perfect storm of chaotic congestion on our highways. In the transition to a fully autonomous world, things could get much worse before they get better.

Figure 107 - In a world where everyone can hail a cheap ride in a self-driving car, the roads may become more congested than ever before—especially on highways that attempt to accommodate *both* autonomous vehicles and human drivers.

Undoubtedly, we're in for a tense time. But none of the obstacles that we have discussed in this chapter are insurmountable. While bicycles can only carry one person, autonomous ridesharing vehicles (by definition) are designed to carry several. Recall that 76.4% of morning commuters are currently driving solo—refusing to carpool to work. So if robo-taxis succeed in coaxing just two people into sharing a ride, then this feat alone could cut morning congestion in half. And, unlike dockless bicycles, autonomous vehicles need not remain stranded at the destination of their last fare. Instead, they can drive themselves back to the highway or park at a suitable waypoint. Additionally, as discussed in Chapter 7, most of our daily errands will someday be fulfilled by autonomous couriers. Many of the reasons that currently prompt people to leave their homes and take to the streets will no longer even exist in the future. The novelty of "riding in a self-driving car" will quickly fade away just as the novelty of "riding in an airplane" has. A legislative and logistical apparatus will evolve to accommodate self-driving technology—just as it has with every other disruptive innovation. It's entirely possible that the majority of our cities transition into the age of autonomy without much more than a couple fender benders.

Still, the bike-sharing companies serve as a valuable analog—illustrative of the unseen consequences that result when nascent transportation technology is hastily deployed in an urban environment. These lessons should not be lost on us. In crafting the autonomous world of tomorrow, we must not solely allow capitalistic impudence to shape our decisions. Inspiriting our cars with self-driving capabilities is only the first step of many. For even if every single car in the country was made capable of autonomous driving today, this feat alone would not endow us with all of the efficiencies of the transportation renaissance. The wheels of these vehicles can only spin at maximum velocity if they ride upon roads that are specifically designed to accommodate their accelerated speeds. Their parcel deliveries can only be efficiently fulfilled when their appendages interact with mechanized receptacles that are capable of receiving their hasty handoffs.

The complete formula for a hyper-efficient autonomous society consists of at least four primary components.

Self-Driving A.I. Technology		Ridesharing Platforms (Robo-Taxis)		Autonomous Courier Vehicles		City Infrastructure Upgrades
	+		+		+	

The clockwork of the autonomous age will only rotate with frictionless efficiency following the integration of *each* of these constituents. In the ontogenesis of this new technology, it is essential that we avoid recapitulating our past indiscretions. Failing to address the problem of transportation *holistically* could result in a tragedy of the commons—our roads made host to a tenuous contest between man and machine; a demolition derby of unruly drivers and unregulated A.I. If we're not careful, then the very machines that were created to *end* congestion may actually become the cause of it.

The Three Most Common Objections to Autonomous Technology

Given the preceding list of logistical, economic, and legal dilemmas, it is clear that self-driving technology will leave future generations with much to grapple with. But, oddly, the above-cited counterarguments are not typical of the vox populi. Instead, most critics are prone to posit more facile objections—three of which we'll consider now.

Objection 1: "Self-driving cars are too dangerous."

About that Self-Driving Uber tragedy...

On the evening of March 18, 2018, Elaine Herzberg was walking her bicycle across Mill Avenue in Tempe, Arizona. At the same time, Uber's modified Volvo XC90 self-driving car prototype was cruising down the northbound lane. At 9:58 p.m. the Volvo slammed into Herzberg and her bike—making her the first pedestrian to ever be killed by a self-driving car.

The most common objections to driverless cars come from those who consider them to be "unsafe." Tragedies like the Herzberg affair make for provocative news headlines that burrow deep into the public consciousness. This case may set a precedent by which future autonomous vehicle incidents will be litigated. And it might go down in history as the first incident of "robot betrayal" on the roads. But let's take a moment to consider the facts:

- The victim (Herzberg) was illegally walking her bike across a four-lane highway—at a point that lacked a crosswalk and proper street lighting.
- The Uber vehicle (being a self-driving prototype) actually did have a human behind the wheel (Rafaela Vasquez) who was tasked with monitoring the car's behavior and asked to "keep her hands on her lap" — ready to seize the wheel if a perilous situation were to arise.

So why did Vasquez fail to stop the car?

According to the 2018 preliminary report by the National Transportation Safety Board:

At 1.3 seconds before impact, the self-driving system determined that an emergency braking maneuver was needed... According to Uber, emergency braking maneuvers are not enabled while the vehicle is under computer control, to reduce the potential for erratic vehicle behavior. [Instead, the] vehicle operator is relied on to intervene and take action.

Uber's self-driving computer *did* detect the presence of an obstacle in the road. However, it wasn't equipped to execute emergency braking procedures. Instead, it relied on its human driver to take over. But Vasquez wasn't paying much attention to the road. Cellphone records would later reveal that she was streaming an episode of NBC's *The Voice* at the time of the collision.

Ironically, this principal tragedy of vehicle automation was actually caused by the most *human* of blunders:

- The victim was jaywalking in the dead of night.
- The driver was watching a video on her iPhone.

If the self-driving car's computer would have been allowed to execute emergency braking functions then the collision may have been avoided.

About that Self-Driving Tesla tragedy...

The second most famous case of A.I. malfeasance happened in Williston, Florida on May 7, 2016. Joshua Brown's Model S Tesla failed to distinguish a white semi-trailer truck from the brightly lit sky. The computer assumed that the road was free of obstructions and proceeded to drive directly *under* the truck's trailer (at 74 miles-per-hour); thus tearing off the Tesla's roof and tragically decapitating its driver. Three years later, Tesla Model 3 owner Beren Banner would suffer an oddly similar fate—colliding with a tractor-trailer on Florida State Highway 441 (just eighty miles east of Joshua Brown's crash site).

The Brown and Banner accidents were widely publicized when they occurred. And they are frequently cited whenever any publication puts out a story about "the dangers of self-driving cars." Several other Tesla incidents have managed to make headlines since that first fateful day in 2016. Given that there are now over 730,000 Teslas (with Autopilot versions 2 and 3) on the road, we are sure to see more of such tragedies and more debates about the safety of self-driving cars.

It is understandable that such stories would arouse public trepidation about sharing our roads with robots. However, this first Autopilot-related death purportedly came after over 130 million miles of uneventful mechanical chauffeuring. When we compare this ratio with the national average (one fatality in every 94 million miles), then Tesla's A.I. driver is (arguably) already safer than a human.

Most importantly, when considering accidents involving Teslas, we should remain cognizant of the number of conventional auto crashes that we *don't* hear about. In 2016 (the year of Joshua Brown's crash) there were an estimated 7.2 million police-reported traffic accidents nationwide—in which 37,461 people lost their lives. (That's one dead American every 14 minutes.) But how many of *these* tragedies made national news?

As time goes by, many people will remember *something* about "that Tesla crash in Florida" and how "the computer killed the driver." But few of us will recall *anything* about the other 37,461 Americans who died in conventional automobiles in the same year.

Better than humans?

For many, the safety features alone will be enough to convince them of the utility of self-driving technology. Recall that, since 1900, our cars have claimed 3.7 million American lives—three times more than our wars have. Even with ongoing research into road design and automotive safety engineering, the average number of US auto fatalities still hovers around 37,000 per year. Recall that 94% of car accidents are the result of human error—the three leading causes being alcohol, speeding, and distraction. Though autonomous vehicles may never be perfect, they at least don't drink, speed, or text their girlfriends while driving.

People are often quick to exclaim how frightened they'd feel with a "robot behind the wheel." But, in light of the above statistics, is it not rational to be even more apprehensive about putting a "human behind the wheel?" Even if you trust the steadiness of your own two hands, how much faith do you have in the hands of your fellow motorists?

- Consider the soccer mom in an SUV full of children—hurtling towards you at 75 miles-per-hour on a two-lane highway.
- Consider the speeding weekend warrior—who just got a text message on his beeping iPhone.
- Consider a car full of teenagers—being teenagers.

How much faith do you have in the driving abilities of these people?

If given the choice, would you rather face off against an oncoming SUV controlled by a human driver (one prone to distraction and who can only attune his senses for brief periods), or would you rather confront a vehicle that features a suite of a dozen sensors and an onboard computer—capable of digesting millions of data points each second?

Even the most ardent opponents of self-driving technology typically agree that driverless cars will (at least someday) be safer than their human counterparts. As their safety record continues to improve, people will eventually be forced to acknowledge the utility of riding in a vehicle laced with a dozen sensors, as opposed to one operated by a bored and easily distracted driver.

Objection 2: "Self-driving cars are an abomination!"

One peculiar objection to self-driving technology comes from those who consider the entire enterprise to be an abomination. For them, careening down the road in a driverless car is akin to tampering with black magic. Or, perhaps more commonly, they consider such cars to be emasculating or dainty—a vehicle best suited for a man who doesn't know how to build his own deck, mow his own lawn, or barbeque. But even if we were to grant that riding in a car driven by a headless horseman is an ungodly act of depravity and sloth, shouldn't one also consider the act of sitting in bumper-to-bumper traffic to be equally abhorrent? What about the daily tragedy of scraping dead people off of the asphalt? Is this ritual not emblematic of the iniquitous state of human transportation?

Curiously, though autonomous vehicles are often perceived as *otherworldly*, they may actually help us return to a more natural human condition. People were never meant to sit with their right foot hovering over an accelerator pedal—isolated in a metal prison and tapping their way home through a sluggish sea of traffic. Instead, isn't the act of reclining in a self-driving vehicle, resting calmly during the journey, and arriving safely at the destination a much more "natural" state of being?

Objection 3: "Driving is a right! And self-driving cars are un-American."

This country has a long-held fondness for automobiles. America's "love affair with the car" is palpably portrayed in our film and literature. Cultural references depicting *freedom* often feature a lone automobile—rolling down a scenic country road or trekking across an arid landscape. Perhaps the emotions manifested by such experiences are not so easily aroused if the vehicle is piloted—not by a rakish rogue—but by the ascetic hand of an indifferent computer. Most auto enthusiasts will not be eager to replace their passion with a computer program. And some will gladly take up arms against those who would deny them of their God-given right to experience fahrvergnügen.

Writing for Jalopnik in 2015, rally race driver Alex Roy described the romance this way:

The car transformed the definition of freedom, annihilating geographic, social and cultural barriers, and came to embody freedom itself. The car has become the object onto which people project their ambitions, and onto which they pin decisions and memories that amplify and reflect who they are... Although technology has begun to dilute our relationship to driving, it cannot dilute our relationship to the

vehicle… Enthusiasts love cars not merely because of what they are, but because they are an extension of the self.

Three years after penning these words, Roy went on to found the *Human Driving Association*. In his 12-point manifesto, Roy outlined his stance against the nascent forces of vehicular automation—particularly those that seek to make all cars into *driverless cars*. Some of his main beliefs are summarized below:

- No vehicle should ever be deployed without a steering wheel.
- No vehicle should be denied access to transportation infrastructure because it is piloted by a human.
- All vehicles must be capable of operating completely independent of any communications network.
- A new constitutional amendment should guarantee the "right to drive" to all Americans.

In the header atop Roy's manifesto sits an image of a steering wheel and a line of copy which references the former slogan of the National Rifle Association. To paraphrase:

"I'll give you my steering wheel when you pry it from my cold, dead hands."

Such rebellious pleas will surely attract driving enthusiasts and car culture fanatics the world over. Organizations like Roy's will undoubtedly grow in numbers—springing up to engage in legislative maneuvering as the armies of autonomous vehicles draw ever nearer.

Car companies seem to be aware of the coming scuffle. In an apparent attempt to assuage the fears of those who like to steer, Tesla CEO Elon Musk took to Twitter in 2015 and wrote:

To be clear, Tesla is strongly in favor of people being allowed to drive their cars and always will be. Hopefully, that is obvious. However, when self-driving cars become safer than

human-driven cars, the public may outlaw the latter. Hopefully not.

Ford CEO Jim Hackett expressed similar sentiments when (in a 2018 interview with Newsweek) he posited that self-driving technology may not be allowed to take over the *entire* vehicle.

It may be that [autonomous] vehicles just keep you from crashing, [but] you still control the way that you're navigating through a city. I think there's something to that, because people like to have a sense of control.

For many car enthusiasts that "sense of control"—that feeling of power, of independence, of the potential for boundless adventure—these emotions were the active ingredients in the formulation of their initial car-buying decision. For them, the value of a car is unrelated to the efficiency by which it carries the driver to his destination. The journey is the destination. If every car comes with a robot bolted to the driver's seat, then… where's the fun in that?

Such forthright deductions may stir the passions of protestors in the coming autonomous disruption. Discussions about "the right to own cars" may eventually become even more polarizing than contemporary quarrels over "the right to own guns."

- The *anti-car lobby* will cite traffic congestion data, pollution problems, and road fatality statistics. And they'll exclaim that cars are just too dangerous to trust in the hands of the average citizen.
- The *pro-car lobby* will insist that it is their constitutional right to operate an automobile on public roads or federal highways.

Such quarrels could devolve into civil unrest. There may be city-wide protests in which the owners of conventional automobiles assemble onto roads that have been cordoned off for "autonomous vehicle use only." They might use their cars to create an impromptu barricade—a damn preventing the sea of self-driving vehicles from proceeding along the highway.

The use of "road jamming" as a form of protest is nothing new. In April of 2019, over 100 truckers came together to make a "slow roll" on Interstate 465 near Indianapolis. Cruising at the minimum highway speed of 45 mph, they made two passes around the circular interstate with their rigs—attempting to call attention to a new federal law that would require Electronic Logging Devices (ELDs) to be installed in all trucks.

Car enthusiasts of the future might resort to similar tactics. In the early years of the driverless revolution, such displays of civil unrest might even be an everyday occurrence. But, eventually, just like every other bolshie protest against the nascent forces of automation, the cries of these protestors will subside—drowned out by the screeching electric blur of exclusively autonomous roadways. Someday, these protesters won't be able to drive alongside the torrent of self-driving cars even if we let them. Their reflexes won't be fast enough. The pitter-patter of their turn signals won't be acknowledged. In the future, watching a human attempt to drive down a road full of autonomous vehicles will look like watching your grandmother attempt to race her 1984 Cadillac Eldorado in the Monaco Grand Prix.

A.I. drivers will courteously tolerate a human presence on the road, at least for a while. They'll relinquish the right-of-way and kindly alter their course to accommodate our momentary lapses of driving etiquette. But eventually, it will become evident that we're just slowing them down. As stories of *human vs. robot* auto accidents begin to hit the daily newsfeeds, the public may initially be inclined to blame the latter. But, as self-driving technology improves, the assignment of blame will shift to the former.

As a 2013 blue paper by Morgan Stanley put it:

Suddenly, the question of "what if I don't want to share the road with *an autonomous car*" could become "what if I don't want to share the road with *someone driving his own car?*"

As the years of the autonomous revolution roll on, young people will eventually grow tired of listening to old people assert their right to gum up the works. Lane by lane, cities will claim ever more roadways for the machines. With each acquisition, the driving skillset will fade into obsolesce. Someday, people who still insist on driving their own cars will be considered quaint or déclassé. The skill of *driving* will continue to be sidelined—eventually settling on par with the likes of sheep herding, trapping, or navel-gazing.

Indeed, this *great un-keying* may already be well underway. In 1983, 46% of American 16-year-olds had a driver's license. But by 2017, that number had dropped to 26%. Though the US population gains two million people each year, Americans seem to be losing an interest in driving. Per-capita miles driven reached their all-time peak back in 2006.

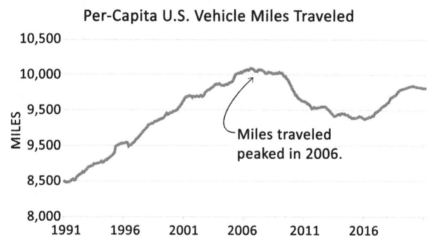

Figure 108 - The number of per-capita miles logged on US roads peaked in 2006. (Source: Data originally compiled by Yonah Freemark of The Transport Politic using data from the U.S. Federal Highways Administration Traffic Volume Trends 2018.)

People don't drive as much as they used to. When autonomous ridesharing catches on, the skill of *driving* will become about as useful to future urbanites as the skill of *horseback riding* is to us now. Just as man has lost the right to ride his horse on the highway, so too will he eventually lose the right to drive his car there. We'll all trade our car keys in for a ridesharing app someday. Though, for some of us, this may not be the most amicable of exchanges.

A City on a Hill

The great disruption that we have outlined in this book is not merely a hypothetical Sword of Damocles. A transitionary period of great social and economic change lies before us. However, we don't know if this paradigm shift will be quick or gradual.

In 1999, the American inventor and futurist Ray Kurzweil proposed his "Law of Accelerating Returns," positing that:

An analysis of the history of technology shows that technological change is exponential, contrary to the common-sense "intuitive linear" view. So we won't experience 100 years of progress in the 21st century — it will be more like 20,000 years of progress... Everyone expects continuous technological progress and the social repercussions that follow. But...few have truly internalized the implications of the fact that the rate of change itself is accelerating.

Perhaps Ray Kurzweil is right. Perhaps the 2020s will present us with a cornucopia of exotic marvels—unveiled at a rate that is multiplicative of that of the 2010s. (Let's keep our fingers crossed.) But given the conspicuous lack of flying cars, robot maids, and space hotels in our lives, it's easy to chuckle at the failures of past prognosticators.

However, when it comes to the deployment of autonomous vehicles, many of the hurdles that lie before us are not of a technical nature. Instead, they primarily are comprised of the vast number of infrastructure upgrades that must be implemented before the age of autonomy can be effectively ushered in. As for the self-driving cars themselves—prototypes have already successfully navigated across the country. The technology needed to drive these vehicles has already been created. Now it is being refined. Eventually, it will be ready for prime time. But nobody knows exactly when or where this unveiling will occur.

Ever the optimist, Elon Musk has long been teasing futurists with titillating copy on his Tesla website for years. Some sentences might lead the reader to believe that Tesla Autopilot functionality is on the verge of achieving its final form. As Verge journalist Andrew Hawkins reported in 2017:

> **Musk promised that once Tesla's Autopilot system had accumulated enough real-world telemetry and data, the company would [put out an] over-the-air software update to all Tesla vehicles [and] enable *full autonomy*. The announcement conjured up a fantastical vision of Musk flipping a switch—to suddenly transform all…Tesla vehicles…into self-driving robots.**

We can imagine a grand gala—with Elon Musk standing adjacent to a gigantic faux red lever. Some midsized municipality has finally volunteered to become "The Nation's First Fully Autonomous City." As the clock tower strikes midnight, the lever will be pulled, the tuxedoed crowd will cheer, and a button marked "ENGAGE FULL AUTONOMY" will illuminate on the user interface of every Tesla owner in town.

Perhaps there will be a control room of Tesla engineers backstage—each one praying that they haven't just conjured up another Y2K. The Tesla PR team will be on deck—attempting to snuff out any reports of panic, chaos, and autonomous destruction (real or imagined). On the following morning, locals could wake to discover their neighbors asleep at their Tesla steering wheels—being chauffeured to work via Elon Musk's newly installed self-driving software.

It could happen that way.

The first several cities to become "autonomous cities" will surely attract media attention and public curiosity. But for most towns, the integration of autonomous vehicles will probably happen in stages—progressing gradually at rates determined by regional demand and legislative prudence.

America is a big place. The urban density of her cities is highly variable and each locale will benefit from automation at differing degrees. Not every neighborhood is stifled with crippling traffic. Not every urban zone is in need of autonomous solutions at the same level of cruciality. The cities that are desperately searching for immediate congestion solutions might serve as our vanguard—the first to tread on the mysterious soil of the autonomous frontier, the first to reap its rewards, and the first to step on its many landmines.

Several nations have expressed interest in enlisting for this service. The sovereign city-state of Singapore occupies only 270 square miles but is home to nearly six million people—making it the second-most densely populated country on the planet. For some perspective, the entire country is *half* the size of Los Angeles—which has two million fewer people.

Given that Singapore is due to gain another one million citizens over the next decade, city planners have been actively searching for solutions to their traffic quagmire. In 2017, Nanyang Technological University (NTU) opened CETRAN—a testing facility for autonomous vehicles featuring a full-scale city mockup.

Figure 109 - The autonomous vehicle test track at Singapore's "Centre of Excellence for Testing & Research of Autonomous Vehicles in Nanyang Technological University" (CETRAN).

This test track contains dozens of obstacles that are commonly found in urban environments—including traffic intersections, bus stops, hills (of varying gradients), and a rain-making machine. NTU's Future Mobility Solutions program director Niels de Boer stated his department's primary objective quite succinctly:

The goal here is to make having your own car completely unnecessary by 2030.

In October of 2019, it was announced that some of Singapore's western roadways will serve as a real world testbed for self-driving shuttles and taxis. The MIT spinoff company Aptiv (formerly nuTonomy) has run pilot robo-taxi programs in Singapore for several years. More recently, Aptiv hooked up with Lyft to offer their robo-taxis in Las Vegas. In May of 2018, a test fleet of 30 self-driving cars was deployed on the strip. And, as of February 2020, the cars have reportedly completed over 100,000 passenger rides.

In a celebratory blog post (that perhaps conveyed even more optimism than the copy on Elon Musk's website), the president of Aptiv (Karl Iagnemma) wrote:

...I'm often asked, "When will self-driving cars arrive?" My answer: "They're already here." If you find yourself in Las Vegas...you can ride in one of Aptiv's self-driving cars, servicing the Las Vegas Strip and downtown area.

Conspicuously missing from Iagnemma's proclamation is the fact that every robo-taxi in Las Vegas must currently come equipped with a human driver—tasked with operating the vehicle manually in parking lots and near hotel entrances. Still, such milestones are worthy of praise. Teaching a robot car to make sense of the chaotic Las Vegas strip (and successfully drive down it without killing anyone) is no easy task. Additionally, Aptiv has built a 130,000-square-foot Vegas research facility—housing an engineering team dedicated to digesting the incoming vehicle performance data and using it to improve upon their ridesharing platform. The lessons learned via such endeavors will be essential in crafting the hardware and software that will enable the genesis of these machines.

Since the *tech companies* own the cars, and the *cities* own the roads, the dawning of the autonomous age can only come about via the collaboration of both parties. Tech companies (like Aptiv) and cities (like Las Vegas and Singapore) must continue to harmonize their interests—venturing hand-in-hand into this new frontier. Striving to expediently (but safely) develop the vehicles and city works that will drive the gears of the coming autonomous revolution. Of all the arguments that we might offer the populace to convince them of the utility of autonomous vehicles, the most persuasive ones will come from showcasing the fruits of autonomy in such test cities. Here, skeptics should be able to ask questions like:

- Did commute times improve following this town's conversion to an autonomous infrastructure?
- What about traffic congestion and air quality?
- Are road fatalities on the decline?
- How fast is Amazon Prime in this area?
- How many minutes does it take to get a pizza delivered?
- How many minutes does it take to get a piano delivered?
- Are the citizens happier now that they aren't required to drive themselves to work each morning?

Observing positive lifestyle changes in cities that have taken the driverless plunge should inspire other locales to consider a similar leap. Perhaps the first such *City upon a Hill* will be founded in Singapore or Las Vegas. (Or in Gothenburg, Madrid, Austin, or Portland.) Either way, a single successful implementation of autonomous vehicle technology will serve as a valuable point of reference. If one city succeeds in creating a thriving autonomous infrastructure then other cities will take notice. And, as self-driving A.I. continues to improve (and driverless tech prices continue to drop) additional regions should be open to investing in similar infrastructure upgrades.

Eventually, every city will jump on the self-driving bandwagon. As this cavalcade picks up the pace, conventional autos will become increasingly sidelined. Municipalities might initially begin their quarantine by creating "no-drive zones"—areas of town where only autonomous vehicles are allowed to roam. Perimeters may be established in urban hotspots like shopping districts, corporate malls, downtown venues, resorts, or college campuses. If conventional auto owners want to access these areas, they may have to park their cars in designated lots on the outskirts of the city, and then ride in on autonomous shuttles. This procedure could serve as a viable transitionary solution. But most drivers will eventually surrender their conventional automobiles in exchange for waypoint-free passage; thus circumventing the need to swap cars at a cordon sanitaire for outmoded gas guzzlers.

Such solutions may appease the car lovers for a while, but they are only delaying the inevitable. All urban centers will be *no-drive zones* someday. The roads will be catering exclusively to robo-taxis and autonomous couriers. In a driverless world, drivers will be *persona non grata*. A holdout car enthusiast will have to satisfy his *need for speed* elsewhere. His craft will become a weekend hobby—delegated to race tracks and country roads. The Mustang he once so proudly paraded on Main Street, soon to be an old gray mare, and no longer allowed in town; put out to pasture in the age of autonomy.

Conclusion

We are blessed with technology that would be indescribable to our forefathers. We have the wherewithal...to feed everybody, clothe everybody, and give every human on Earth a chance... [We] now have the option for all humanity to make it successfully on this planet in this lifetime. Whether it is to be Utopia or Oblivion will be a touch-and-go relay race right up to the final moment.

– Critical Path
by Buckminster Fuller (1981)

That Which is Seen

In this book, I have attempted to present a positive case for the adoption of self-driving vehicle technology while warning readers of the turbulent disruption that will come in the wake of its launch. The voyage to Autopia may be long and perilous but a bountiful reward will go to the generation that manages to successfully navigate to her shores.

In an effort to provide you with a roadmap for this journey, I've summarized its six most momentous milestones below. As you scan the horizon of the future, keep an eye out for each one.

MILESTONE 1

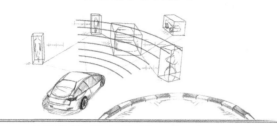

SELF-DRIVING A.I.
SOFTWARE IS PERFECTED

A.I. researchers will succeed in creating self-driving vehicle prototypes that are capable of "Level 5" autonomy—meaning that the car can be built without a steering wheel and can navigate on any road that a trained human driver can manage. When this skill threshold can be reliably achieved then the autonomous revolution will be well underway. But self-driving passenger cars may not be the first successful implementation of this technology—trucks will.

MILESTONE 2

THE DAWN OF
AUTOMATED FREIGHT

Autonomous semi-trailer truck prototypes have already successfully logged thousands of driving miles across America's highways—albeit with a human minder sitting in the driver's seat. Soon, unmanned trucks will be tested on interstates and uneventful roads. Given that long-haul freight transport often takes place over lengthy stretches of tame highway—these routes will serve as a suitable proving ground for self-driving A.I. Moreover, by familiarizing people to this technology (via peripheral exposure), the public will eventually become accustomed to cruising alongside driverless vehicles. If these trucks can keep their safety record unblemished, then citizens should be receptive to proposals that would call for the integration of self-driving cars into their communities.

MILESTONE 3

ROBO-TAXI FLEETS ARE
DEPLOYED IN EVERY CITY

Self-driving A.I. will eventually graduate from *hauling freight* and move on to *hauling people*. Initially, personal autonomous vehicles (of the sort that Tesla has been teasing for years) will be sold like conventional automobiles—purchased, maintained, and used by a single owner-operator and permitted to function autonomously on designated roads. But this transportation model will be short-lived. Large fleets of ridesharing robo-taxis will be mass-produced and deployed in every major US city. Personal car ownership will be rendered uneconomical and most commuters will trade their car keys in for a ridesharing app. The populace will finally be free from the yoke of monthly car payments, gas station visits, driving tests, tire changes, oil changes, break pad changes, and tense negotiations with used car salesmen. The service of transportation will be provided at a price that is far lower than the annual cost of $9,282 that Americans are currently paying to fuel, license, finance, and maintain their cars. As the technology needed to manufacture and operate robot-taxis comes down in price, so too will the cost of a trip. Someday, when the commoditization of urban transport is complete, robo-taxi fares may be so low that they are quite nearly nominal.

MILESTONE 4

PASSENGER DROPZONES
REPLACE PARKING LOTS

Autonomous ridesharing vehicles don't park at the passenger's destination. They drop riders off at the building's curbside. As these cars become the preferred mode of transportation, commercial buildings will need to be affixed with reception areas in which to receive the incoming fleets of robo-taxis and couriers. At first, these zones may merely be demarcated by a painted strip of sidewalk. But in time, all apartment complexes, hotels, shopping centers, and office buildings will be equipped with dedicated facilities designed to receive robo-taxis and their passengers. When the migration from personal car ownership to autonomous ridesharing is complete then the vast asphalt deserts that surround our buildings will no longer be of any use. City planners and developers will be able to repurpose or eliminate 90% of the one billion parking spaces in America.

MILESTONE 5

ALL PARCELS ARE
DELIVERED AUTONOMOUSLY

As retail continues to migrate to the web, an ever-expanding network of delivery trucks and couriers will be needed to fulfill customer orders. Currently, after an online purchase is made (e.g. on websites like Amazon,

Walmart, or Wayfair), the gears of a grand order-fulfillment clockwork spin into action. A picker retrieves your product from a warehouse, hands it off to a packer, who in turn boxes it up and hands it off to a human delivery driver. But someday, there will be no humans involved in this process at all. Autonomous couriers will be used to deliver almost all retail purchases. Your incoming parcels of pet toys and paintbrushes will find their way from a warehouse to your front door without ever being handled by a single person.

MILESTONE 6

HOMES ARE REDESIGNED FOR THE AGE OF AUTONOMY

Our homes were never designed to receive incoming packages at the voluminous quantities that will be possible in the age of autonomy. Due to the many logistical inefficiencies of residential delivery, this *last mile* of order fulfillment is the most challenging leg of a parcel's journey. Currently, for each box in a delivery driver's vehicle, he is required to locate the front door of the destination address, find a temporary parking spot for his truck, carry the box to the front door, and hope that someone is home to receive it. If our houses are to swiftly interact with the autonomous courier vehicles of the future then this procedure will need an upgrade. Homes and apartment buildings will be retrofitted with mechanized dropboxes designed to automatically receive incoming deliveries. When every porch in the country is upgraded in this fashion, the hyper-efficiencies of the autonomous age will finally be available to us. Every purchased good will be delivered by automatons. Our roads will be teeming with self-driving courier vehicles— diligently scurrying across the highway, solely devoted to the pursuit of rapid order-fulfillment. When this day comes, the channels that span between buyer and seller will widen—enabling torrential levels of trading volume to surge forth; the nation will be cast into a mercantile frenzy that has never been seen before. Most importantly, life's necessities will be available for purchase at prices that are lower than they have been at any point in human history.

The Value of Time

When the above six enchantments are finally conjured, their synergy will (hopefully) bring about a world of marvelous conveniences—one in which parcels and people are effortlessly transported with miraculous efficiency. Aside from these direct benefits, there is yet another commodity that is to be gained via the implementation of this technology—*time*. Autonomous vehicles will allow for a reduction in the amount of time it takes to: commute to work, to shop, to complete errands, and to perform common household duties. Homemaking and parenting in the 21st century often requires a mother to wear many hats. Aside from her typical maternal responsibilities, she has traditionally been cast in the role of taxi driver, personal shopper, and chef. But thanks to autonomous vehicles, the amount of time required to complete such chores will be reduced.

- **Taxi** services (like Uber and Lyft) currently do not allow passengers under the age of eighteen to ride without an adult. But future ridesharing platforms will have no such restrictions. Tomorrow's children will begin hailing robo-taxis shortly after they receive their first cellphone—thus eliminating the need for their parents to invest in a minivan and devote so many hours to hauling packs of wild children around town.

- **Shopping** duties (buying groceries, clothing, school supplies, etc.) will continue their migration to online retail channels. The goods and services that were traditionally obtained via daily milk runs will instead be ordered via internet and delivered by autonomous couriers.

- **Home cooking** tasks will continue to be sidelined. In 1965, American women spent an average of two hours per day in the kitchen. But by 2016 that number had fallen to under an hour. When the industrialization of meal production and the automation of home delivery are complete then hungry family members will just summon or reheat meals as needed. The "home cooking" skill will be deprecated—becoming just another quaint hobby like gardening or basket weaving.

Our ancestors spent most of their time worrying about their next meal and guarding hearth and home. But homes of the future won't be built with a hearth. They won't need much more than a kitchenette. Thanks to autonomous vehicles, the three preceding labors that parents have endured (especially since the advent of the automobile) are soon to be outsourced to the machines. As UBS observed in their 2018 paper *Is the Kitchen Dead*:

A century ago, many families [had to produce] their own clothes... The cost of purchasing pre-made clothes from merchants was prohibitively expensive... [Eventually,] industrialization increased production capacity and costs fell. Supply chains were established and mass consumption followed. Some of the same characteristics are at play here: we could be at the first stage of industrializing meal production and delivery.

This next step in the ongoing division of labor will greatly reduce the amount of time needed to accomplish housekeeping and parenting responsibilities. The principal skillset of the average American is about to be atomized yet again.

- Cooking will be industrialized.
- Transportation will be commoditized.
- *All* products will be purchased online.
- And *everything* will be delivered autonomously.

When such a grand autonomous infrastructure finally exists (one capable of transporting children and cheesecake across town in minutes), then the very nature of *housekeeping* will change. Take a moment to consider the amount of time that future families will *not* be spending on present-day tasks.

- Time *not* spent sitting in traffic.

349

- Time *not* spent shopping (for groceries, gifts, household essentials, school supplies, etc.).

- Time *not* spent transporting family members (to school, playdates, soccer, or piano practice).

- Time *not* spent picking up meals, purchasing food, nor preparing it for dinner.

Truly, if *anyone* can appreciate the value of an innovation capable of such time-saving efficiencies, a parent can. The above-listed tasks are representative of the majority of their parenting efforts. Currently, the completion of these chores requires the dutiful navigation of a vehicle through a series of waypoints. But the machines of the autonomous age will be transporting our cargo for us—thus eliminating the necessity of our own participation in these errands and gifting each of us with hundreds of salvaged man-hours per year.

How nice it will be to get this time back. Recall that the average American spends about four years of his life in a car. The vast majority of these outings are solo undertakings. Our freeways are filled with millions of lonely souls— just trying to move their bodies from the place they are, to the place they'd like to be. Just trying to get home. For too long, the service of transportation has only been provided following the payment of a mandatory toll— moments of sentience that must be sacrificed to the gods of navigation. A fare that we have reluctantly handed over—only because there was no other way to get from here to there.

But soon there will be.

In the coming age of autonomy, we will no longer be forced to expend our attention on the monotonous chore of daily commuting or benumbed travel. We need not endure one of life's colorless chapters for the promise of a livelier one. Instead, our time spent in a car will be ours again. The story of our lives will continue to develop whilst we're en route.

As the CTO of *NEXT Future Transportation* Tommaso Gecchelin put it:

I imagine a future where the traffic will be so light that we can [reclaim] the lanes, and put green spaces [there]. A future where bicycles and pedestrians can live together in equilibrium. I think about the future, I dream about the future—where traveling will *not* be more wasted time, but it will be *life in motion.*

That Which is Not Seen

While this promise of a "life in motion" is a captivating dream, it currently remains just that. As with any work of futurology, we look through a glass darkly. Just as we cannot know which of these wishes will come true, so too can we not anticipate their unintended consequences—which will only be revealed after the autonomous rubber hits the road.

At this juncture, the one thing we can be sure of is that our existing transportation infrastructure is inadequate. It will eventually fail us. The old roads that form the hatchwork of our cities were never designed to handle the speed and volume of 21st century commerce. If we are to navigate out of this logistical quagmire then we must seek out an upgrade to this antiquated system. At the moment, self-driving A.I. appears to be the solution we've been waiting for.

The gift of autonomous transport will enable one man to send an object to another man—without the need for a third man's participation in the process. At first glance, this feat may not seem momentous. But if you take a moment to contemplate the nature of this task (i.e. the mere act of moving something from here to there), you'll realize that it actually constitutes the majority of human efforts.

Clearly, this technology will soon change the way we think about movement and the way we design our transportation corridors. But it will also change the way we design our lives and our culture—altering our conception of words like "community," "neighborhood," and "home," and prompting us to reexamine the primal forces that have simultaneously bound us together and pulled us apart since the dawn of civilization.

Good fences make good neighbors

When societies form, individual families are tasked with the chore of partitioning the space that will separate their home from the commons.

- We desire to reside in proximity to other people, but we don't want them to live *too* close to us.
- We yearn to be accepted in a community, while also claiming private ground for a space of our own.

Harmonizing such divergent imperatives is the source of much interpersonal strife. Robert Frost's 1914 poem "Mending Wall" nicely describes this pickle. In the first line, the speaker remarks:

Something there is that doesn't love a wall.

And he wonders why his neighbor compels him to mend the stone wall that divides their properties.

He is all pine and I am apple orchard.
My apple trees will never get across
and eat the cones under his pines...
...
Before I built a wall I'd ask to know
what I was walling in or walling out,
and to whom I was like to give offense.

Building the wall seems pointless. And yet, his requests for justification are repeatedly countered by his neighbor's refrain:

Good fences make good neighbors.

Soon, the machines of the autonomous age will enable us to transcend the walls of time and space that have traditionally helped us reconcile such tensions. With these vehicles, we will traverse the peaks and valleys that divide us—indeed have coexisted with us—since the origin of the species. But once this technology is perfected, our achievement may be a Pyrrhic victory—for we may have underestimated the importance of the space that lies between us.

It could be that the gaps that separated us were also keeping us together. Perhaps these chasms (these natural perimeters) are an essential component in the formation of a unified community—just as the skin is required to hold the body together or as the walls of the castle secure the welfare of the keep. Given that the west's adoption of multiculturalism seems to be resulting in increasingly brittle societies, this is not a difficult thesis to grasp.

The efficiency by which future goods and labor will be transported could expose these communities to a new suite of economic tensions—e.g. increased labor competition, economic interdependence, and foreign market intrusion. When worker commuting routes are fully automated, employers will be able to tap into a labor supply that exists hundreds of miles away from the actual job site—possibly in a different country. In this new age of effortless travel, citizens may be forced to mingle in multicultural environments that they are not accustomed to; navigating the emerging social landscape will be made all the more complex thanks to the rapid infusion of differing norms, races, classes, and creeds.

Some communities may recoil when they realize the ease with which autonomous vehicles will enable "less desirable" citizens (people from "the other side of the tracks") to have free and casual access to Main Street. Just as corporate buildings feature elevators that refuse admission to employees who do not hold an executive key, some future American cities may not be readily accessible to *all* Americans. Neighborhoods might utilize *no-drive zones* to refuse admission to robo-taxis carrying "undesirable" passengers.

Many methods could be employed to conduct such an evaluation. Some cities might simply deny entrance to any passenger who couldn't provide proof of residency. Other cities might adopt a more forensic approach. Suppose you get in a car and ask the computer to drop you off in the "nice part of town." Facial recognition software may be employed to determine your identity. A rapid background search could scour the internet in an attempt to assess your character. This process might tap into several national or private databases—asking questions like:

- Do you have an outstanding arrest warrant?
- How much money do you currently have in your bank account?
- What is your credit score?
- Did you complete high school? College?
- Do you currently take any medication for mental health problems?
- Have you ever been diagnosed with Tuberculosis, Typhus, HIV, or COVID-19?
- Did you say anything politically incorrect on Twitter last night?
- Are you planning on attending the protest that is currently going on near your requested destination address?

Alternatively, a future image processing A.I. may be able to assess your character by merely viewing the taxi's incoming video feed.

- Is your appearance or disposition similar to others who have been recently convicted of crimes in the area?
- Do you appear drunk?

- Do you appear to be carrying any weapons?
- What's your current temperature?
- And are you dressed "properly" for the venue?

Such inquiries might be conducted in a fraction of a second. With the blink of a scanner, the car might decide that you're simply not worthy of a ride to the requested destination. You'll then be asked to choose a different destination or vacate the vehicle.

In considering the preceding, it's easy to envision a dystopian future—one in which *all* cars are self-driving cars and *all* destinations have a criteria threshold that each passenger must meet before access is granted to the region. If Charles Murray is correct (that America is about to undergo a period of unprecedented socioeconomic stratification), then autonomous vehicles might be just the tool that the elites have been waiting for. A scalpel—able to excise the undesirable plebs from the manicured streets of the aristocracy. Such a system will give a whole new meaning to the term "gated community."

If this glimpse of the future makes your skin crawl then you're not alone. *"Something there is that doesn't love a wall."*

Autonomous vehicles will provide us with effortless travel, extended leisure time, and burgers-on-demand. But the unseen costs of these favors may be too great. These machines may usher in a divisive age—fraught with social strife along cultural, class, and racial lines. We may be ensnared in a Faustian bargain by which we will acquire the gift of omnipresence but lose our nation's soul in the process. In our quest to knock down the geographical barriers that divide us, we might create a nation that is more divided than ever before. The future tapestry of America may be woven with an impressive transportation network, but stained with large splotches of urban blight and dotted with the exclusionary walled gardens of the elite.

Revolutions are messy affairs

There is a risk that autonomous technology will be used to cordon off some of our more dilapidated urban environments—further dividing the populace and stressing the social and economic rifts that already run so deep between us. But unearthing the edges of this strata is not a difficult job for anyone willing to dig a little. And given recent demographic changes, the bifurcation of these fissures is likely to progress anyway.

In 1990, America's cities were home to 250 million people. By 2100, they will need to accommodate twice that number. We *will* see an increase in our present-day urban problems: traffic congestion, traffic fatalities, last-mile delivery issues, a lack of affordable housing, food deserts, smog, homelessness... All of these issues will compound as the population continues to grow. Given that the commoditization of transportation via autonomous vehicles has the potential to curtail (or even eliminate) most of these issues, we have to (at least) give them a try. Besides, at this point, we're running out of options.

The difficulties we endure in contemporary urban environments will continue to stack up. Like blocks in a game of Tetris, each new obstacle is the result of the accumulation of generations of hastily improvised solutions—stemming back to the Machine Age (1880 to 1945) and sometimes much farther. Our current strategy to increase the throughput of the roads of mechanization has been to construct ever *more* roads and ever *more* mechanical vehicles. But the benefit of this patchwork is transient. In most US cities, congestion is worse than ever. New city corridors fill up with cars just as soon as the asphalt dries. We need a better way to move people and parcels around town. At this point, self-driving A.I. seems to be the best candidate for the job.

While the *age of autonomy* may be a period of great disruption, it need not be a violent disruption. While it is possible that there will be moments of pandemonium in some parts of the country, it is just as likely that (for most of us) the transition to an autonomous infrastructure will be uneventful and gradual. So much so that, by the time we're ready to trade our car keys in for a ridesharing app, self-driving vehicles will be old hat. The coming revolution might turn out to be about as radical as the rebellion we all lived through two decades ago—when we switched from VHS to DVD.

In any case, we are wise not to rush the process. Patience will be required of us in the years to come. Most of these technologies will take decades to implement; not necessarily because of the inadequate pace of self-driving A.I., but rather, because of limitations in the speed by which we'll be able to incorporate these vehicles into the existing urban infrastructure and legislative framework.

We've got a long road to haul. But thankfully, a new generation of scientists, engineers, and legislators has risen to the challenge. It will be up to them to build the autonomous world that has been described in these pages. It will be up to them to decide how their communities will accommodate the coming cavalry of headless horsemen, and to determine the extent to which they will be allowed to roll through the city gates. The generation that will be tasked with enduring the pangs of this revolution has already been born. As for the generation after that, they won't recall ever riding in a car with a steering wheel.

Is this the road to Cockaigne?

This is not the first book to herald our imminent arrival to the Land of Plenty—aka Elysium, Erewhon, El Dorado, or Eden. Many men have cited nascent technological advances and proclaimed (with great exuberance) that techno-utopia is to be found just around the next bend—e.g. Karl Marx, John Maynard Keynes, Oscar Wilde, John Stuart Mill, Edward Bellamy, and just about every single science fiction writer in the grave.

So what's the difference between us and them?

Well, they didn't have artificial intelligence and autonomous vehicles.

We do.

So there.

Truly, many important scientific achievements have been won in the last fifty years. The wonders that 21st-century technology is capable of producing are far superior to anything that could have been conjured up in the days of the Utopian Socialists. So, in the least, this generation has a *better* chance of achieving techno-utopia than the previous one. And I do think that A.I.-driven autonomous technology is one of the tools that will help to get us there. But this is not the crux of my argument.

The construction of an autonomous infrastructure will not necessarily result in a state of techno-utopia. A state of dystopia is just as probable. Like any other instrument of science, these vehicles can be used for the betterment of man or the opposite. They won't solve all of our problems. However, though they may not delineate the final rung on the ladder of progress, they are (in the least) high enough to lift us up out of the muck.

We climb now in the second decade of the 21st century, yet we have not managed to rise much above the ground. The legs of our ladder sit in a primordial ooze. Though we have managed to traverse a few rungs, the swamp gas is still quite odorful from our current position. For even in this supposedly enlightened era, the vast majority of us are still struggling to fulfill life's most elemental needs.

Recall that, for the average American, 61.6% of monthly expenditures are directed to just three items: *transportation, food,* and *housing.* But thanks to the coming autonomous renaissance, we may soon reach a point where these three expenses are substantially reduced:

1. **Transportation** hardware for self-driving vehicles will initially come at a high cost—only available as an optional upgrade in pricey sedans. But its extravagance and exclusivity will be short-lived. Like all technology, the price of self-driving A.I. will fall precipitously after its initial release. And it will keep falling until, someday, the commoditization of transportation will be complete and the cost to ride in a robo-taxi may not be much more than a present-day bus fare.

2. **Food** production costs will be greatly reduced thanks to an autonomous supply chain. And, thanks to autonomous couriers,

individual meals can be mass-produced in large food-preparation facilities and delivered to the front doors of individual consumers as needed. In time, the price of a hot meal should be much cheaper than anything currently available at a café or fast-food restaurant.

3. **Housing** of the future can be constructed far outside of the city center. Thanks to high-speed autonomous ridesharing vehicles, millions of outlying acres will someday be in range of jobs and services. People will be able to rent or purchase homes on the outskirts—where land is plentiful and cheaper. Additionally, since parking lots will not be needed in the future, all of these precious urban acres can be repurposed or converted into yet *more* residential units—thus increasing the housing supply and further reducing the cost.

While self-driving vehicles might not immediately deliver us to utopia, the three gifts that they predictably have in tow are quite nearly as grand. If the men of the mind manage to successfully engineer an autonomous infrastructure, then their achievement will free mankind of the three heaviest burdens that he has beared since the advent of civilization—*shelter*, *transport*, and *food*.

- Truly, *this* is the most noble and vital implementation of all the technology that we have discussed in this book.
- This is the point of our conversation.
- This will be mankind's reward for centuries of scientific labor.

Figure 110 - Hephaestus in his forgery. A portion of the salon ceiling in the Barberini Palace. Fresco by Pietro da Cortona (1639).

Leaving the Dark Ages of Transportation

It should be clear to you now that autonomous vehicles are much more than just "cool cars." Conventional automobiles are not imbued with the spirits of autonomy. They have no brain with which to navigate; they have no eyes with which to see the road. But soon they will have both. And they will use their digital senses to lead us out of this dim era. Atop their backs we will sit, glance over our shoulders, and wave goodbye to another milestone of the industrial revolution—thus delineating the end of the dark ages of transportation.

Our inability to solve the *riddle of logistics* has kept us in the shadows for far too long. Autonomous vehicles are carrying us to the point of this puzzle's completion. Like excited children, many are scrambling to set the last few jigsaw pieces into place. With wide-eyed anticipation, we await the reveal of the assembled image. We yearn to delight in the sublime coherence of its final form. And we hope that the completed mosaic depicts a picture of a better world.

The zeitgeist of the 21st century is so often a salty soup. But lately, it is sometimes served up with a dash of optimism. I can taste it.

I can feel it.

Can you?

Fait Accompli

Now we have come to the end of the book and it's time for you to offer your opinion.

So what do you think? What do you think about self-driving technology? Do you think that autonomous vehicles will be good or bad for our civilization?

You may still be suspicious of these machines.

- Perhaps you have doubts about their safety.

- Perhaps you don't think that America will be able to successfully transition to an autonomous economy.
- Perhaps you think the cons outweigh the pros.
- Perhaps you think that these machines are driving us right over a cliff—where we'll crash and burn in a dystopian fireball of misery and chaos.

Either way, I hope that this book has managed to prepare you for some of the questions that you are soon to encounter on your journey through the 21ˢᵗ century. But whatever your opinion might be, it is with regret that I must reveal the following news:

Your opinion doesn't matter.

And nor does the opinion of Tucker Carlson, Ben Shapiro, Andrew Yang, Elon Musk, John Maynard Keynes, or the POTUS. Because, regardless of the *type* of world that results from the use of autonomous vehicles, the citizens of *that* future world will be unable to fathom an age without these machines.

- They will watch old movies featuring the brown skies of smoggy Los Angeles (or Delhi or Beijing or Peshawar), and they will recoil in disgust and pity.
- They will cringe when considering the tedium of bumper-to-bumper traffic and the boredom of being a passenger in a vehicle without internet connectivity.
- They will look at auto fatality statistics or watch car crash videos on YouTube and they will become wide-eyed with horror. We'll explain to them that, in the past, road trips would take place on two-lane highways catering to human-driven cars—each one hurtling toward the other at 65 miles-per-hour—with only a line of yellow paint to divide them. Future generations will regard such journeys as *suicide missions*.
- They'll look back on us, shake their heads and wonder, "How could anyone live like that?"

Our children's children will consider debates about the utility of autonomous vehicle technology (like the conversation between Ben Shapiro and Tucker Carlson) to be ridiculous. They'll experience the same degree of befuddlement that we would experience when watching a video of two commentators from the 1980's debating the pros and cons of owning a cellphone. There's *nothing* you could say to present-day Americans that would convince them to give up their cellphones. Even if you could somehow guarantee that their annual household income would instantly *double* at the very moment that each citizen relinquished their iPhone, this incentive would *not* be enough to get most of them to do it.

Similarly, citizens of the future will not defenestrate their robot chauffeurs in exchange for a fairytale about sustained economic growth or job stability. Autonomous vehicles will be as fundamental to their civilization as water and sewer lines. A child born today will soon grow accustomed to reaching for his phone and tapping his finger to summon taxis, tacos, and toothpaste. Future children will begin hailing robo-taxis shortly after they learn how to tie their own shoes.

Figure 111 - A young child boards an autonomous school bus.

The events of their lives will playout whilst they sit atop these machines. Autonomous ridesharing platforms will carry them to their first rock concert, their first high school dance, and their first funeral. In their history classes, they will learn that, in the early years of the 21st century, some politicians attempted to curtail job-loss by hindering the deployment of autonomous technology. They'll consider such legislative ploys to be delusional.

Indeed, even if we entertain a Bizarro World thought experiment—in which a nation-state succeeded in banning autonomous vehicles outright, this implementation of Neo-Luddism would only succeed in making the region unable to compete on a global scale. If the goods and services of *Nation A* are produced via a supply chain that operates, say, 25% more efficiently (thanks to AV technology), and *Nation B* has the *same* production capacity but refuses to use autonomous vehicles, then *Nation A* will beat *Nation B*. The global markets will not respond kindly to countries with a technophobic proletariat.

For good or bad, right or wrong, the swarm of autonomous vehicles is headed our way. They will buzz along the roads of our neighborhoods, dispense their cargo into the receptacles of our residences, and pervade casually into our lives—with the same perfunctory ease by which bumblebees integrate into our gardens. But unlike the "Africanized killer bees" of 1980s fame, this swarm is real; let us begin to prepare for it with wisdom and prudence. We need not react with the hysterics of the Luddites—smashing these machines, setting them ablaze, and cursing them for their wicked efficiency—as if their wheels were enlivened by witchcraft and their batteries were fueled by treachery. Instead, let us courageously walk into the theater of the new world. Let us hope that, when we peel back the curtain of the future, a new stage of abundance awaits.

As the cost of transporting people and parcels approaches zero, man's ability to manipulate the location of matter will no longer be so intractably constrained. Thus far in our existence, this most elemental of maneuvers has only been achieved following the payment of a hefty toll. But we will soon be free of this exigent debt. The men of science have punched the ticket for us.

Undoubtedly, these machines will improve the efficiency by which we deliver *sustenance to people*. But, perhaps more importantly, they will reduce the cost of delivering *people to people*. Thanks to this newfound flexibility of travel, autonomous vehicles may allow for more harmonious relationships—easing the burden of bringing families together and facilitating interpersonal connections with an efficacy that rivals the synthetic interaction we currently attain via cellphone or internet technology. In the future, we need not so often settle for mere digital communication in lieu of tactile exchanges. Instead, we can opt for face-to-face encounters—thus enabling the transmission of the ineffable subtleties of discourse which are only conveyed via our presence and our touch. *"Showing up is 80 percent of life."*

Over a century ago, E.M. Forster understood the limitations of interaction via electronic devices. In 1909, almost nobody in Britain owned a landline telephone. But Forster envisioned a time in which every home would be equipped with an "optic plate"—a handheld screen with a built-in camera that enabled people to communicate via both audio and video. Today, the name for this technology is "videotelephony" but initiating such conversations with PCs or cellphones (via Skype or Zoom or Apple FaceTime) is so commonplace that we have little use for the technical term.

In *The Machine Stops*, the face of the protagonist's son (Kuno) appears on the optic plate. Desperately, he pleads with his mother to visit him—not merely through the faux imagery created by *The Machine*, but in person, in real life. Lamenting the inadequacies of remote communication, Kuno tells her:

The Machine is much, but it is not everything. I see something like you in this plate, but I do not see you. I hear something like you through this telephone, but I do not hear you. That is why I want you to come. Pay me a visit, so that we can meet face-to-face, and talk about the hopes that are in my mind.

Perhaps this innate desire—for proximity to the ones we love the most—will finally be adequately facilitated by the mechanical marvels of the coming autonomous age.

Pay me a visit, so that we can meet face-to-face, and talk about the hopes that are in my mind.

The Machine Stops by E. M. Forster (1909)

Journey well.

Made in the USA
Las Vegas, NV
28 September 2022

56152783R00213